Cities Without Capitalism

This book explores the interconnections between urbanisation and capitalism to examine the current condition of cities due to capitalism. It brings together interdisciplinary insights from leading academics, activists, and researchers to envision progressive, anti-capitalist changes for the future of cities.

The exploitative nature of capitalist urbanisation, as seen in the manifestation of modern cities, has threatened and affected life on the Earth in unprecedented ways. This book unravels these threats to ecosystems and biodiversity and addresses the widening gap between the rich and the poor. It considers the future impacts of the capitalist urbanisation on the planet and the generations to come and offers directions to imagine and build de-capitalised and de-urbanised cities to promote environmental sustainability. Written in lucid style, the chapters in the book illustrate the current situation of capitalist urbanisation and expose how it exploits and consumes the planet. It also looks at alternative habitat practices of building autonomous and ecological human settlements, and how these can lead to a transformation of capitalist urbanisation.

The book also includes current debates on COVID-19 pandemic to consider post-pandemic challenges in envisioning a de-capitalised, eco-friendly society in the immediate future. It will be useful for academics and professionals in the fields of sociology, urban planning and design, and urban studies.

Hossein Sadri, an academic, activist, architect, permaculture designer, and urbanist, is teaching contemporary challenges in architecture at Coventry University and Girne American University. His fields of research include de-urbanisation, future cities, and design of socio-ecological and ethical habitats.

Senem Zeybekoglu, an academic, architect, passivhaus designer, permaculture designer, co-founder of De-Urban Design Studio and director of Ecodemia: An Ethical and Socio-Ecological Academia, is a full professor at Girne American University.

Routledge Studies in Urbanism and the City

For more information about this series, please visit https://www.routledge.com/
Routledge-Studies-in-Urbanism-and-the-City/book-series/RSUC

Cities Without Capitalism

Edited by
Hossein Sadri and Senem Zeybekoglu

Foreword by
Peter Marcuse

Routledge
Taylor & Francis Group

LONDON AND NEW YORK

First published 2022
by Routledge
2 Park Square, Milton Park, Abingdon, Oxon OX14 4RN

and by Routledge
605 Third Avenue, New York, NY 10158

Routledge is an imprint of the Taylor & Francis Group, an informa business

British Library Cataloguing-in-Publication Data
A catalogue record for this book is available from the British Library

Library of Congress Cataloging-in-Publication Data
A catalog record for this book has been requested

ISBN: 978-0-367-37060-2 (hbk)
ISBN: 978-1-032-04306-7 (pbk)
ISBN: 978-0-429-35248-5 (ebk)

Typeset in Times New Roman
by Apex CoVantage, LLC

To our Future: Aras and Ege, and all the children
of the world

Contents

Contributors

Tom Angotti is Professor Emeritus of Urban Policy and Planning at Hunter College and the Graduate Center, City University of New York. He was the founder and director of the Hunter College Center for Community Planning and Development. His recent books include *Transformative Planning: Radical Alternatives to Neoliberal Urbanism, Zoned Out! Race, Displacement and City Planning in New York City, Urban Latin America: Inequalities and Neoliberal Reforms, The New Century of the Metropolis, New York For Sale: Community Planning Confronts Global Real Estate*, which won the Davidoff Book Award, and *Accidental Warriors and Battlefield Myths*. He is an editor of progressivecity.net and Participating Editor for *Latin American Perspectives* and *Local Environment*. He is active in community and environmental issues in New York City.

Juliana Birnbaum trained as a cultural anthropologist and skilled in four languages, and has lived and worked in the United States, Europe, Japan, Nepal, Costa Rica, and Brazil. She is the co-author of two books Sustainable [R]evolution: Permaculture in Ecovillages, Urban Farms and Communities Worldwide, (2014) and CBD: A Patient's Guide to Medicinal Cannabis (2017). She is also the mother of two daughters and has attended over 100 births as a doula and assistant midwife. Juliana currently coordinates the Volunteer and Faculty departments at Spirit Rock Meditation Center in Woodacre, CA.

Mary H. Dellenbaugh-Losse is an urban researcher, consultant, and author focusing on the development of equitable, co-produced cities. Her work spans a diverse range of topics, including urban commons, cultural and creative industries, intermediate and adaptive reuse of vacant buildings, participation of underrepresented groups, and the political, symbolic, and normative aspects of architecture and urban planning. She lives and works in Berlin.

William W. Goldsmith, professor emeritus, Cornell University, studies US segregation, inequality and poverty, and international urbanisation. Goldsmith taught in Puerto Rico, Colombia, Brazil, Italy, and at UC Berkeley. He helped found Planners Network and the Planners of Color Interest Group (POCIG), and served on the Governing Board of ACSP. Born in San Francisco, he studied at UC Berkeley and Cornell.

Kanishka Goonewardena is an Associate Professor in the Department of Geography and Planning at the University of Toronto. His writings explore the intersections of urban studies and critical theory, with an emphasis on the works of Henri Lefebvre, Fredric Jameson, and anti-colonial writers.

David Gouverneur holds MArch in Urban Design from Harvard University and BArch from Universidad Simón Bolívar in Caracas. He is an Associate Professor of Practice in the Departments of Landscape Architecture and City Planning at the Stuart Weitzman School of Design at the University of Pennsylvania. He has been Chair of the School of Architecture at Universidad Simón Bolívar, co-founder of the Urban Design Program and Director of the Mayor's Institute at Universidad Metropolitana in Caracas, and Adjunct Secretary of Urban Development of Venezuela. He identifies himself as one of close to six million Venezuelans forced to leave the country, due to ideological fanatism, totalitarianism, economic meltdown, ethical and institutional decomposition, violence, and neglect of the territorial–urban discourse.

Andrew Heben is an urban planner and designer, and the author of *Tent City Urbanism: From Self-Organized Camps to Tiny House Villages* (2014). Many of the ideas within his book have since been put into action through the co-founding of SquareOne Villages, a nonprofit organisation creating self-managed communities of cost-effective tiny homes for people in need.

Terry Irwin has been a design practitioner for more than 40 years and was a founding partner of the international design firm MetaDesign. From 2009 to 2019, she was the Head of the School of Design at Carnegie Mellon University (CMU), Pittsburgh. She holds an MFA from the Allgemeine Kunstgewerbeschule, Basel (1986), Switzerland and an MSc in Holistic Science from Schumacher College/Plymouth University (2004). She is currently a professor in the School of Design and is the Director of the Transition Design Institute at CMU.

Gideon Kossoff is a social ecologist and environmental activist with a PhD in design from the Center for the Study of Natural Design at the University of Dundee, Scotland (2011). He is currently the Associate Director of the Transition Design Institute at Carnegie Mellon University as well as a member of faculty in the School of Design and an advisor in its doctoral program in Transition Design. His research focuses on how transforming the relationships between humans, the natural environment, and built worlds can become the foundation for a sustainable society. His doctoral thesis is summarised in a chapter in the book "Grow Small, Think Beautiful", edited by Stephan Harding.

Peter Marcuse, a planner and lawyer, is Professor Emeritus of Urban Planning at Columbia University in New York City. His fields of research include city planning, housing, homelessness, the use of public space, the right to the city, social justice in the city, and globalisation. He has several books including but not limited to *Searching for the Just City*, with multiple co-editors, and *Cities* *for People, Not for Profit*, co-edited with Neil Brenner and Margit Mayer, published by Routledge.

Porus D. Olpadwala is Professor Emeritus of City and Regional Planning and Dean Emeritus of the College of Architecture, Art and Planning of Cornell University, and Adjunct Professor in the College of Architecture and Planning of the University of New Mexico. His areas of scholarly interest are comparative political economy, economic development, and urban and environmental issues.

Sinead Petrasek is a doctoral student in the Department of Geography and Planning at the University of Toronto. Her research and practice centres around art, urban studies, and social inequality. She is currently designing a project on social reproduction and co-operative housing in Toronto.

Jakob Rigi is an associated researcher with the Frontlines of Value project. He holds a PhD from the department of Anthropology and Sociology, SOAS, University of London. He has taught at Cornell University and Central European University and been a fellow at NYU, Edinburgh University, Manchester University, and SOAS.

Yavor Tarinski is the author of the book "Direct Democracy: Context, Society, Individuality" (Durty Books 2019). He is co-founder of the Greek libertarian journal Aftoleksi, member of the administrative board of the Transnational Institute of Social Ecology, as well as bibliographer at Agora International. In Bulgaria he has co-founded social centre "Adelante" – the first self-organized social centre in the country – as well as the first Bulgarian Social Forum.

Hossein Sadri, an academic, activist, architect, permaculture designer, and urbanist, is teaching contemporary challenges in architecture at Coventry University and Girne American University. His fields of research include de-urbanisation, future cities, and design of socio-ecological and ethical habitats.

Senem Zeybekoglu, an academic, architect, passivhaus designer, permaculture designer, co-founder of De-Urban Design Studio and director of Ecodemia: An Ethical and Socio-Ecological Academia, is a full professor of Architecture at Girne American University.

Figures

Tables

Foreword

Cities after capitalisms, cities after viruses

This book could not have appeared at a more fortuitous time. It addresses issues raised by the COVID-19 but will continue to be troublesome after the virus has been conquered, and need to be faced in a post-survival world, while there is still time to influence what that world should be like.

Written before COVID-19 had become the centre of all-consuming attention, the book addresses topics that are matters of deep concern today, in questions confronting us already today in fighting that virus, not so much how to survive it but so much on what after survival has been assured.

Many issues arise from considering cities without capitalism, confront alike the relationship of cities with capitalism and with economic development in the fight against the coronavirus. The issues can be formulated as this book does, as the tensions between cities and capitalism, or as the tension between economic concerns of capitalism and the humanistic concerns of urbanism and cities.

Tensions appear in discussions of how much of the multi-trillion survival fund agreed upon by the Congress should be devoted to supporting the economy, a capitalist economy, as opposed to how much should to meeting the costs of medical treatments and saving all lives at all costs, the humanitarian focus, physically concentrated in cities. What we learn here is that the tension between the economic/capitalist and the humanitarian/urban is not a new tension, but structural in our society. Today it is a tension around responses to the viral pandemic, but it has deep historical roots. It is a tension that will likely survive well past the life course of the COVID-19 and beyond differences about the desirable role and shape of cities.

Climate change, raised in a number of chapters here, reflects much the same historical tension. The length of the perspective in which it is addressed by public policy; who is which position in assessing the seriousness of the problem, whether a matter of human survival or not; how the sides line up on each issue with the interest groups contending in the political arena, as Trump versus Sanders (to oversimplify) as this is written; these questions parallel the one posed here.

The tension between the two sides is today almost measurable in dollars and cents. Is it more important to spend a trillion dollars in sustaining our economy, our material, and technological productivity, or is our goal a social one, to deepen our solidarity with each other as social beings and enrich our cultural lives? Which comes first, the economic or the social? Are they mutually dependent, or alternatives? Capitalists presumably would measure success in dollars and the accretion of wealth, urbanists in lives saved and the quality of life in cities, for the rich and the poor alike, capitalists perhaps less so.

Unavoidably, a second related theme is discussed by contributors to this book. It is simple, but profound: What is it that we want to have survived the coronavirus: our wealth or our humanity? What capital provides, or what cities provide? Are the two ultimately mutually interdependent, or are they ultimately inconsistent and indeed contradictory alternatives between which we must choose? Is wealth, and its pursuit (not just in capitalism) necessary for our cities to thrive and their cultural values to flourish, or is pursuit of wealth itself antithetical to the flourishing of humanist values, of a popular humanism such as many of the chapters describe?

The contributions that follow illuminate another aspect of the similarities between the long-term capitalism/cities' tension and the short-term issues the pandemic has engendered. It is the question of power and democracy. At stake are indeed central questions about our values, our priorities for our future as a species. Let cities stand as centres nourishing humanitarian values, our social goals, our social relationships, and let capitalism stand for the organisations and purposes of economic activities, providing the material infrastructure for all of our activities, consistent with some values, inconsistent with others. How are we to achieve the most desirable balance between the two? Is that a utopian idea, or are there suggestions as to how we get from here to there? Is democratic decision-making, inefficient as it may be, better than autocratic decisions, ignoring or manipulating the decisions of the majority?

And who will answer that question? Most of the chapters that follow come out on the side of democracy and would agree that cities are the physical and social forms that best serve democracy, and see capitalism as largely coming out on the other side. There is much history presented that illuminates the conflict, although it is inevitable.

In addressing the future, the question is first and foremost, survival itself. Everything needs to be done quickly. Decisions need to be made. Issues of power need to be resolved. A trillion dollars need to be spent. Should they go to resuming maximum growth or to revitalising the classic cultural values of cities? Assuming we survive the virus and the internecine strife over values and goals so graphically described in some of the chapters, the question of what comes next needs to be debated even under the pressure of time. Such debate is no longer the luxury it may seem to be. If it waits until survival is no longer in doubt, the opportunities to mold the future in the most desirable directions will be badly diminished. The chapters in this book, written before the corona pandemic, push the concern with the future beyond survival. An optimistic viewpoint, perhaps, today. But addressing the post-survival future explicitly assumes, there is a light at the end of the tunnel. Strength can be drawn from imagining what that post-survival future might look like. Optimism is not in rich supply. Maybe the chapters in this book can help strengthen the resolve of those engaged in the struggles involved.

Peter Marcuse

Introduction

Hossein Sadri and Senem Zeybekoglu

Urbanisation grew in the interstices of cities and shortly spread towards rural and wild areas to radically transform not only the feudal order, but virtually all biotic and abiotic beings on the planet and their historical and paradoxically ecological relations. Capitalism and urbanisation drew their first breath simultaneously, twins who suckled the same she-wolf. They aged together, embraced and reproduced a new exploitative system with all of its residual inequalities, injustices, vulnerabilities, violations, degradations, extinctions, and even biological mutations. Sweeping away the Earth's creations and diminishing biodiversity, this murderous duo became the greatest man-made hazard.

Capitalism and its accompanying urbanisation have been the most important factors affecting most aspects of our lives and the future of our planet in the last two centuries, causing an ecological crisis that adds up to several other humanitarian crises such as poverty, hunger, forced migrations, and wars. The capitalist urbanisation not only sets the rules as to how humans should live, but also endangers wildlife, ecosystems, forests, rivers, seas, biodiversity, and native communities all around the world, changing our planet in ways often irreversible. If the urbanisation that stems out of this grow-or-die economic mindset continues, the next generations will inherit a more polluted, unjust world with scarce resources.

The end of this road is visible; either we humans eventually overturn this threat or it could spell the end of humanity. At this crossroad, it's absolutely necessary to stitch the fragmented relations between all oppressed beings, from starving children to tortured fighters, from deserts to oceans.

If we want to save our planet, home of our species and innumerable others, we have to take immediate action and strongly object to cruel enforcements of the established order, sold to us as the only way to continue our "normal" lives within our comfort bubbles. If we want cities to be humanitarian and ecological habitats, and hope for a world freed from capitalism and urbanisation, we must first envision it, implement it, build it with our hands before more species go extinct, more people starve to death, more waste is piled up on the planet's surface and in outer space. We need to free science, art, and philosophy from the bridle of power relations, and utilise these human faculties to rehabilitate our relations with nature and other human beings.

In order to create a better world, we need to imagine de-capitalised and de-urbanised cities that nurture, regenerate, enrich, enable, and support ecosystems and biodiversity, produce soil and food, harvest water and sun, and freed from waste and fossil fuels. We need to imagine equitable, inclusive, just, egalitarian, solidarist, non-hegemonic, and non-hierarchical cities that provide endless opportunities to their human inhabitants so that they can discover and develop their abilities in order to contribute to the well-being of our species and our planet. We need to imagine cities without capitalism.

**

The quest for a world freed from capitalism and urbanisation with thriving de-capitalised and de-urbanised cities is this book's main motivation. During our 2018 sabbatical in the City College of New York as invitees of the late urbanist, critic, and activist, our comrade professor Michael Sorkin, we were highly focused on the idea of "De-Urban Design", trying to solidify this concept through an in-depth exploration of New York City. Evolving from a small town into the heart of global capitalism, the city inherently possessed a blend of wealth and deprivation, opportunities and obstacles, struggle and fraud. In our commute to the city, oppression was all too palpable; we could almost breathe poverty and misery on the streets. On the other hand, we observed with hope and excitement the grassroots movements, protests, resistance, and solidarity practices taking place in neighbourhood meetings, community gardens, streets, and public spaces which gained momentum with the Occupy movement.

The foundations of this book, among many other projects, were laid during this short sabbatical period. Sorkin's essay "Architecture without Capitalism" featured in Peggy Deamer's book "Architecture and Capitalism" ignited the idea of the title. The project aimed to bring together academics, activists, and researchers from different disciplines to probe into relationships between urbanisation and capitalism in order to outline the existent condition of cities, address anti-capitalist and anti-urbanist transitions, and envision progressive alternatives that could shape the future of cities.

With this aim in mind, the book is structured under three main sections.

The first section of this book is entitled "Cities and Capitalism". It seeks answers to the question of "Why capitalist urbanisation cannot sustain?" in order to expose the inherent conflicts of capitalist urbanisation by revealing the unsustainability in the relationship between urbanisation and capitalism. This section contains four chapters.

The first one, "Cities without capital", investigates systemic ways of interaction between urbanisation and capital by using Karl Marx's laws of movement and is written by Prof. Porus D. Olpadwala, the former Dean of Cornell University.

The second chapter, "Cities and subjectivity within and against capitalism", is written by Prof. Kanishka Goonewardena and Sinead Petrasek, two scholars from the University of Toronto and discusses the contradictory modes of human fulfillment and negation in urban habitats.

The third chapter entitled "Can urbanization reduce inequality and limit climate change?" underlines the unsustainability of current urbanisation patterns, environmentally and socially and is written by Prof. William W. Goldsmith, Emeritus Professor of Cornell University and the author of Saving U.S. Cities.

Andrew Heben, the author of the "Tent city urbanism", wrote the fourth and last chapter of the first section, addressing the unwinnable battle for space in the capitalist city especially by the unhoused.

The second section of the book is entitled "Cities against capitalism". It also contains four chapters, but here the main question is "how the alternative habitat praxises of today can motivate the transition of capitalist urbanisation". This section aims to illustrate current praxises and alternative movements of building more autonomous, communal, and ecological human settlements against capitalist urbanisation.

The first chapter is written by the pioneers of the idea of the "transition design", two scholars from Carnegie Mellon University, Dr. Gideon Kossoff and Prof. Terry Irwin. The chapter is entitled "Transition design as a strategy for addressing urban wicked problems" and covers a wide range of challenges that current urbanisation faces and the possibility of creating a catalysing societal transition towards more sustainable and desirable long-term futures.

The second chapter of this section is entitled "Transition pioneers" and discusses the cultural currents and movements of our time that "preveal" the future post-capitalist city. This chapter is penned by Juliana Birnbaum, the co-author of "Sustainable [r]evolution".

The third chapter is written under the title of "Urban commons: toward a better understanding of the potentials and pitfalls of self-organized projects" by Dr. Mary H. Dellenbaugh-Losse, co-author of "Urban commons: moving beyond state and market". This chapter reveals how commons projects present a possible "third way" to manage urban resources in the niche between the market and the state.

The fourth and last chapter of this section is written by Prof. David Gouverneur of the University of Pennsylvania. The chapter's title is "Counteracting the negative effects of real estate-driven urbanism + empowering the self-constructed city" and addresses social inclusion and environmental soundness, with a hybrid approach, merging planning and design with the dynamism of the informal city.

The third and the last section of this book is entitled "Cities without capitalism". The three chapters in this section deal with the question of "what a non-capitalist city will look like". This section introduces progressive visions related to cities of a post-capitalist era.

The opening chapter is written by Prof. Tom Angotti, an emeritus professor of CUNY and the author of several books including "Transformative Planning: Radical Alternatives to Neoliberal Urbanism" and "Zoned Out! Race, Displacement and City Planning in New York City". Under the title "What will a non-capitalist city look like?" this chapter investigates the diverse challenges of the non-capitalist city and the ways it can overcome them.

The second chapter "Towards democratic and ecological cities" is drawn up by Yavor Tarinski from the Transnational Institute of Social Ecology. It puts emphasis on the grassroots dynamism of possible future sustainable cities from the perspective of social ecology.

The closing chapter is written by Dr. Jakob Rigi, anthropologist and former faculty of Cornell University and Central European University. Entitled "The coming revolution of peer production and the synthetisation of the urban and rural: the solution of the contradiction between city and the country", it manifests the spatial logic of the replacement of capitalism with the Commons-Based Peer Production.

The foreword of this book is written by one of the main pioneering urban thinkers of our time, Prof. Peter Marcuse, emeritus professor of Columbia University. Referring to the COVID-19 pandemic, the foreword is entitled "Cities after capitalisms, cities after viruses".

1 Cities without capital

A systemic approach

Porus D. Olpadwala

Introduction

The question of cities in a post-capitalist world may be addressed in two ways. Since we do not know what the new world will look like, we investigate capital's impact on urbanization to identify features that the new society might wish to retain or discard. The second is to examine the experience of cities in the few post-capitalist societies that had a transitory existence in the last century.

Our entry point into capitalist urbanization will be Marx's laws of motion or movement. These include *inter alia* growth, commoditization, economic concentration, technological dynamism, internationalization, unemployment, economic and social inequality, recurrent crises, and financialization. Marx formulated these to capture the intrinsic or inherent features of capital defining all aspects of the system. We will use them to study capital's relationship with urbanization.

We take the tendencies as given, but derive them anew using everyday business propositions instead of the labor theory of value, in hopes of better associating them with contemporary modes of thought (Section II). Next, we consider the practical advantages of incorporating a systemic viewpoint in urban analysis (Section III). A case study follows in applying the tendencies of capital approach to an urban issue, in this case, housing (Section IV). The following section (Section V) considers shortfalls in understanding that can occur when systemic elements are not included. Section VI is a short post-script about post-capitalist urban experience.

Capital

Capitalist enterprise is subject to two mostly immutable laws. One is the drive for *private, unlimited, financial, accumulation*. The second is *unceasing competition*. These are imperatives and not choices. Each italicized item is critical, with substantive change in any negating capital's governing conditions. For example, if the goal is not accumulation, or the accumulation is required to be limited, or it is intended for public and not private good, or if the competition were not all-encompassing, it would not be capitalism as we have known it.

The rules for accumulation are minimal and direct. They require that self-interest drive all decisions, that costs are externalized wherever possible while gains are made internally, and that immediate gratification be preferred to longer-term benefit – the principles of the invisible hand, maximization of returns, and the time value of resources, respectively.

The goal of business is to make money in unlimited quantities. There exists, however, substantial scope in how this can be done as circumstances and preferences intervene to take the process in separate ways. This amply is borne out in capital's multifaceted history of operating through trade, conquest, production, and usury. However, the nucleus of the system has remained intact even as site and context have metamorphosed and it is possible to deduce a few economic and social *general tendencies* for capitalist societies.

Table 1 explores the influence of business motivations and behaviors on capital's tendencies. Production for private gain creates a dissonance between private return and public well-being to the detriment of the latter, for example, in the promotion of harmful products like cigarettes. The dissonance favors social inequality as in the pharmaceutical industry choosing to address the less acute but more profitable demands of the better-off over more urgent but less remunerative needs of the poor. Another result is the constant push for the cost externalizations to shift liabilities from the private to the public sphere, for instance, in dealing with pollution remediation, or by not providing a living wage such that even full-time employees (the working poor) need to be supported from public sources. All these behaviors are benefited by larger size, thus creating also a tendency towards market power and concentration (1.1.a–d). The imperative of unrelenting competition reinforces these socially harmful tendencies and compels firms to operate at the level of the lowest common denominator (1.2). The push of unlimited financial accumulation is an important driver of growth and concentration as there exists no internal system limit to growth, indeed, it is just the opposite (1.3).

These motivations generate a series of business behaviors (1.4–1.11). Making money requires that revenues be maximized (1.4–1.8) and costs minimized (1.9–1.11). To enhance revenues, firms must try to sell more of their existing products at home and abroad (1.4), bring new products to market (1.5), increase sales prices (1.6), speed the turnover of capital, that is, minimize the time between making investments and capturing returns (1.7), and maximize firm influence over public authorities, for example, to reduce regulations and taxes, etc. (1.8).

To lower costs, firms must endeavor constantly to reduce the quantity of inputs, in both labor and material (1.9), pay less for what they use (1.10), and, again, maximize public influence, in this case to reduce external obligations such as duties and taxes (1.11).

The table tracks in the second column the impact of each behavior on the formation of capital's tendencies. For example, the need to expand market share (1.4), increase selling prices (1.6), lower input prices (1.10), and maximize public influence (1.8 and 1.11) results in a tendency towards *economic concentration*. The requirement to keep creating new products (1.5), speed turnover (1.7), and decrease the quantity of inputs (1.9) promotes *technological dynamism*, while market expansion (1.4) and the need to reduce the price of inputs (1.10) stimulates *internationalization*. (For a detailed derivation of the tendencies with empirical examples, see Olpadwala 2019.)

Table 2 collects and groups the tendencies shown in the second column of Table 1. It shows that a social system based on privately owned resources used for unlimited private accumulation de-linked from social need tends to veer towards increased *consumption and growth (2.1), commodification in most aspects of life (2.2), economic concentration (2.3), technological dynamism (2.4), and internationalization in outlook and operation (2.5)*. Such a system generates simultaneously *employment* (2.6) and *unemployment* (2.7) that net regularly in the unemployment that is an important contributor to social *inequality* (2.8). Finally, the system is prone to economic imbalances that generate regular *financial and economic crises* (2.9).[1]

Capital and a systemic approach to urbanization

Urbanization theories fall into two nested categories, the general and the particular. General schemes analyze urban development using universalist principles of economies of scale, agglomeration economies, and comparative advantage that focus on the minimization of production and transportation costs without regard to historical context.[2] The analysis of the particular requires supplementing the general with explicit consideration of the essential goals and mores of the society in question. For capital, this means making tendencies of movement a central part of urban analysis.

Gordon's 1976 study of two centuries of US urbanization illustrates the importance of this duality. He tracked the growth and spread of towns and cities and their internal allocation of spaces for work, residence, and recreation. In Gordon's schema, the general laws of scale and agglomeration constitute the *quantitative efficiencies* of urbanization and by themselves are not sufficient to explain all location decisions. A more complete understanding needs also to include systemic elements, in this case pertaining to the capitalist labor process. Business places a premium on access to pliant labor and these *qualitative efficiencies* are as and sometimes more critical for profit and growth. Location decisions therefore are an amalgam of the two. A site with both types of efficiencies trumps those with only one. Where competing sites have comparable quantitative efficiencies, the decision goes to the one with greater qualitative strengths. In extreme cases, businesses even are willing to sacrifice short-term quantitative efficiencies to consolidate a long-term control over labor.

Hymer's work on the internationalization of capital offers another window into systemic thinking (1979a, 1979b). Location theory works with geographic and size distributions of towns and cities, the sittings and hierarchies of urban growth. Early Western[3] treatments focused on agriculture and the countryside, with transport differentials between town and country counted as the key variable determining spatial outcomes (von Thunen 1966). With the growing importance of production in the economy, attention shifted to the division of labor and scale and agglomeration economies in creating the spatial canvas. Geographer economists Christaller (1966), Losch (1954), and Weber's (1929) stylized spatial hierarchies are some of the better known examples. Less known, but more relevant because they are less abstract and rooted in empirical observation, are Adam Smith's studies of urbanization and Britain's urban hierarchies (Stull 1986).[4]

Urban hierarchies based on the division of labor differentiate between settlements on the basis of size as defined by the number of occupations within a village, town, or city. Smith posited that the division of labor determines the size of a market, hence the greater the number of occupations, the larger the city. This, however, is a static representation that does not account for the forces behind any particular distribution of occupations. Hymer addresses that by adding, as Gordon does, the qualitative efficiencies of control so important to business to the quantitative efficiencies of the division of labor to produce a dynamic urban hierarchy. He finds "a correspondence principle" whereby the vertical division of labor that centralizes control within corporations reproduces itself as a hierarchical division of labor between cities and regions nationally and internationally. "Status, income, authority, consumption [would] radiate out from these centers along a declining curve, and the existing pattern of inequality . . . would be perpetuated."[5] This fusion of the hierarchies of power and division of labor provides a more realistic approach to understanding location decisions.

Harvey (1985) is arguably the best-known proponent of capital's unique, symbiotic relationship with urbanization.[6] While towns and cities have contributed to many other types of societies, earlier urbanizations were ancillary to growth whereas Harvey considers urbanization critical to capital. He argues for a vital and reciprocating bond between the two, a connection so essential to capital's health that it makes urbanization a systemic force in its own right.

Harvey views urbanization serving capital by mobilizing, appropriating, producing, and absorbing economic surpluses.[7] Tributary and slave societies like Egypt and Rome contained a handful of cities that contributed three of the four functions, with production mostly absent. Tribute, not trade, linked them with their hinterlands, and surpluses were mobilized and appropriated in a parasitic as opposed to symbiotic relationship. The advent of manufacture in Medieval towns added the fourth element of production, and the guilds and charters of early Mercantile capital sought out towns and cities to gain shop-floor economies of scale, larger agglomeration economies in location, and access to trade. This started the process of making cities important to business but it was only the ascendance of joint stock companies with their imperative of unlimited financial accumulation that made urbanization indispensable to capital.

The pursuit of boundless growth puts increasing pressure on firms to secure mounting scale and agglomeration economies available mainly in cities and as a result deepens the bond with and dependence on urbanization. Businesses might like to be perceived as "rugged individual(s) – diligent, intelligent, and above all frugal"[8] who create wealth but as Beauregard (2018:25) points out, "(U)ntil public wealth is put in place, cities do not even begin to generate private wealth." The larger the scale and complexity of business, the more reliant capital has to be on the public largesse of cities.[9]

For Harvey both bond and dependence go even deeper. A system of billions of people making innumerable daily decisions about production and consumption geared to "accumulation for accumulation's sake, production for production's sake" (p. 129) based solely on market prices inevitably slides into recurring gridlock with goods remaining unsold, workers unemployed, and profits lacking channels for re-investment. Harvey reasons that capital's antidote over the last century and a half for tackling surplus product, labor, and capital has been massive investment in physical infrastructure centered on the urban-built environment.

> (T)he production of built environments became servants of capital accumulation . . . [with] the physical and social landscape of capitalism increasingly caught up in the search for solutions to the over-accumulation problem through the absorption of capital and labor surpluses by some mix of temporal and geographical displacement of surplus capital into the production of physical and social infrastructures.
>
> (pp. 195–196)

The corrective is a public–private exercise in which both sides contribute financial resources and develop innovative financial institutions and instruments. Harvey (2008) reviews the positive impact of debt-financed investment in infrastructure and property on capital's health, starting with France in the mid-19th century where massive infrastructure investments at home and abroad, most notably the Suez Canal and Haussmann's reconfiguration of Paris, enabled by new financial institutions like Crédit Mobilier and new debt instruments, ameliorated a synchronous crisis of surplus capital and surplus labor. The pattern of heavy investment in the built environment coupled with continuous financial innovation continues through this day with new forms of securitization like packaging local mortgages, CDOs (credit default options), and subprime mortgages spreading risk globally and enabling an "astonishing integration . . . of debt finance urban development all over the world."[10]

Urban analysis benefits from incorporating elements of the particular, embodied by the internal logic of the relevant social system, with generic or universalist approaches of economies of scale, agglomeration economies, and comparative advantage. We turn next to how the internal logic of capital as something by its tendencies of movement relates to urbanization.

Capital's tendencies and urbanization: housing as a case study

Urban aspirations center on good livelihoods, housing, transportation, health, sanitation, education, recreation, and community. We use housing as a case study to track capital's tendencies in practice. The question to ponder is whether the state of housing in the United States is the expected result of capital's genuine tendencies or whether it is evidence of a distorted, compromised, or corrupted capitalism.

On any given night in 2018, approximately 553,000 people in the United States were homeless. Two-thirds had some form of limited and transient shelter but a full third survived on "streets, abandoned buildings and other places not suitable for human habitation" (HUD 2018; National Alliance 2019). This is a point estimate, so the actual number of homeless people would be much higher as individuals cycle in and out of the condition. In New York City in 2019, for example, 62,321 people used municipal shelters on any given night, but over the year that number was 133,284 (Coalition for the Homeless 2020).

Young people comprised a significant portion of those affected. In school year 2017–18, 1.5 million students reported being homeless at some point in the previous three years. This represented a 15 percent increase over two years. Seventy-four percent were in shared housing, 12 percent in shelters, 7 percent in hotels and motels, and 7 percent were unsheltered. The unsheltered cohort had increased by 137 percent over a three-year period (National Center 2020).

The United States had 138.5 million housing units in mid-2018 (U.S. Census Bureau 2020a). At the same time (second quarter 2018) the Federal Home Loan Mortgage Corporation estimated that the nation had a shortfall of between one million and four million units (FreddieMac Insight 2018b).[11]

The shortfall is concentrated at the lower and lowest ends of the income scale. Better placed households had 7.4 million second homes in 2016 that accounted for 5.6 percent of the total housing stock (NAHB 2018), plus they owned other categories of real estate (U.S. Census Bureau 2020b, Vacancy Rates). Nearly 200 cities had a median home value of at least $1 million as of June 2018 and 23 were projected to join them in 2019 (Loudenback 2019). At the stratospheric end, Fortune reports that in 2016, 540 US billionaires owned an *average* of 9 homes (and 19 cars) *outside* the United States (Shen 2016). Domestic details on numbers of homes are not available but Forbes identified 33 properties listed at $75 million or more in 2017 and 18 more at $100 million (Sharf 2017).

Cross-border real-estate investment works both ways. Thompson 2020 notes that New York City developers placed a "huge bet" on foreign plutocrats including "Russian oligarchs, Chinese moghuls and Saudi royalty." They were joined by billionaires from Australia, Hong Kong, Turkey, and India in snapping up "second (or seventh)" homes in Manhattan.

Many of these homes, including those of domestic tycoons, remain unoccupied for much of the time. Solomon 2019 references a NYC Department of Housing Preservation and Development study stating the number of pieds-a-terre jumping from 55,000 to 75,000 between 2014 and 2017, with 5.400 units worth $5 million plus. A third of 92,000 condos in 773 buildings did not claim primary residence on Department of Finance tax rolls and the US Census Bureau deemed 60 percent of residences in a 14-block tract of Manhattan to be "seasonally vacant." US Representative Carolyn Maloney points out that in her Manhattan East Side District there are complete buildings where there are no lights on and calls them mere "bank accounts." "Not since the Gilded Age . . . will so much expensive real estate be so little inhabited" (USA Today 2013).[12]

The shortfall in housing units puts financial pressure on renters as well as putative homeowners. The country lost four million low-rent apartments between 2011 and 2017 (Cohen 2019) and New York City alone lost one million between 2005 and 2017 (Honan 2018). Over the same period, rentals charging $1,000 or more rose by five million, those charging between $600 and $999 dropped by 450,000, and low-cost rentals under $600 fell by 3.1 million. The share of low-cost units in national rentals fell from 33 percent in 2012 to 25 percent in 2017 (Joint Center Harvard 2020).

Builders are expected to complete some 371,000 new apartments in 2020, compared to 247,000 in 2019 and 119,000 in 2010 but in many metropolitan areas between 60 and 89 percent of the apartments will be in higher-than-average rent neighborhoods (Editorial Board 2020). As the supply of low-cost housing dwindled, households with $75 thousand incomes accounted for three-quarters of growth in renters (3.2 million) between 2010 and 2018 while those earning less than $30 thousand fell by nearly one million (Joint Center Harvard 2020).

Eighteen million severely burdened households managed to keep a roof over their heads only by paying more than 50 percent of their income on housing. They spent on average 37 percent less on food and 77 percent less on healthcare than households in housing that they could afford ($700 on average for all non-housing costs including $310 on food). Moderately cost burdened households for whom rent was between 30 and 50 percent of total expenditure spent 13 percent less on food and 40 percent less on healthcare (Joint Center Harvard 2019).

The travails of the rent burdened reach into other aspects of life. For example, they are forced into longer commutes. Four million people spend at least three hours driving to and from work each day, or more than 90 minutes each way. The proportion of these so-called supercommuters has grown to 2.8 percent of all commutes since 2005 and is especially high around cities "with booming economies and big housing shortages." In the San Francisco metro area, supercommuters more than doubled from 2005 to 2016 and in Seattle rose by 66 percent (Holder 2018).

The ultimate agony is eviction. Princeton's Eviction Lab notes that the narrative on housing focuses on homeowners but "there is an ongoing epidemic of eviction and displacement in the renting market." From 2000 to 2016, 1 in 17 renters were served notices and 1 in 40 actually were evicted. At the peak of financial crisis in 2010, slightly over one million home foreclosures were completed nationally but almost a million tenants are evicted *every single year*" (Eviction Lab Updates 2018, emphasis added).

A variety of government programs provide affordable homes for roughly five million people but fall far short of the 18 million noted earlier who also need help. Two deep-subsidy assistance programs of the Department of Housing and Urban Development (HUD) have retrenched in the recent past with the number of households in public housing dropping from 1.14 million in 1993 to 1.02 million in 2016, and privately owned subsidized projects declined from 1.72 million in 1993 to 1.37 million in 2016. A third program providing housing vouchers increased throughout the two decades to reach a total of 2.30 million (Kingsley 2017).

Adding injury to insult, much of the nation's public housing stock is in poor condition with broken boilers, mold, mildew, rodent infestations, and fires creating hazards, injuring residents, displacing people, and destroying units (Urban Wire 2020). For example, from 1 October 2017 through 22 January 2018 more than 80 percent of New York City's Housing Authority residents (143,000 apartments, 323,098 people) were without heat for an average outage of 48 hours (Neuman 2018). In New York City and across the nation federal divestment in public housing has led to a repair backlog of $70 billion and $32 billion, respectively (Velazquez 2019).

Housing and capital's tendencies of movement

How is it that wealthy cities and countries are unable to properly house their inhabitants, now to include also substantial numbers of the middle classes (Dewan 2014). Even the most die-hard apologists for the *status quo* would be hard-pressed to argue for an ignorance of the problems or a paucity of resources to solve them. That leaves an intermediate spectrum stretching from indifference at one end to caring but ineffective remedial efforts at the other. The treasure expended on public social programs[13] together with the extensive time and resources contributed to private charity make it clear that it is not indifference. Thus the question is why people of good will, with knowledge and resources to spare, are unable to prevent such deprivation. Working through capital's tendencies as they interact with housing provides one source of clarification.

For capital, housing is a *commodity* like any other, to be produced and sold for private gain.[14]

No surprise therefore that where supply is concerned, developers and renters focus on the high end of the market, exacerbating the *inequality* that already exists. The move is abetted by growing *concentration* among housebuilders resulting from the failure of hundreds of smaller units during the Great Recession (Timiraos 2015). The home-rental market remains fragmented, but the nearly $25 trillion combined value of US homes has made it one of the biggest asset classes in the world and attracted Wall Street's attention. The world's largest private equity firm, Blackstone Group, and other Wall Street titans bought $40 billion worth of deeply discounted foreclosed homes by the thousands after the mortgage meltdown, exacerbating both *commoditization* and *concentration* in rentals (Dezember 2017).[15] *Inequality* also has restricted supply as better-off owners agitate against additional taxes and higher-density housing, particularly lower-cost homes and apartments, and cash-hungry cities opt for commercial zoning to generate sales taxes. At all levels, there exists extensive *internationalization* with significant cross-border participation in investments and deals.

On the demand side, tendencies towards *concentration*, *technological dynamism*, and *internationalization* put steady downward pressure on jobs and wages, making *unemployment* and *inequality* a staple even in good times and impairing prospects for securing decent housing. The United States' "market poverty" rate remained roughly unchanged from 27.0 percent in 1967 to 28.7 percent in 2012.

Two-thirds of the market-poor or 19.2 percent of the entire population were in "deep poverty" meaning they exist on less than half of the poverty line (White House Data).

This approach suggests that persistent housing problems for poorer people are inbuilt into capitalist operations. Their existence is not evidence of a failure in the system but rather confirmation that capital is functioning as expected.

Capital's history demonstrates that it produces social problems in tandem with economic wealth at rates that wealth seemingly is unable or unwilling to address.[16] Businesses must focus on narrow self-interest and the proverbial bottom line to survive, and such social benefits as accrued from their working mostly are the indirect results of profit seeking (Olpadwala and Mansury 2007). Public policy interventions to alleviate poverty are more direct, but they too must accommodate bottom-line demands. Thus affordable housing increasingly has become the preserve of developers and landlords, public health of insurance and pharmaceutical companies, nutrition programs of the agricultural sector, and international assistance closely aligned with geopolitical interest.

Because the essence of the problem lies in its nature as much as its magnitude, many potentially effective housing alleviation measures are not easily implementable, or if they are, they are not replicable or scalable. It is a matter of running hard to move slowly. This is not to dismiss or minimize the importance of these efforts. But it is to stress the importance of systematic thinking in urban analysis and policy.[17]

A-historic nature of much contemporary urban analysis

Unfortunately, this is not the case much of the time. Capital's tendencies are as firmly entrenched as ever – indeed, Einstein's observation that "nowhere have we really overcome . . . 'the predatory phase of human development'" (Einstein 1949) – is even truer now, but capital's inner workings still are not made endogenous to most social analyses. For example, economist Paul Krugman asks "Why Can't We Get Cities Right?" in an op-ed stating "(I)t's not hard to see what we should be doing . . . however, in practice, policy all too often ends up being captured by interest groups" (Krugman 2017). The interest groups have been around and dominant for decades yet it is rare for mainstream analysts to investigate the systemic forces underlying them.

One result of this is limited or partial explanation. Bettencourt (2013) stresses that a major drawback in urban analysis is that problems are "treated as independent issues" and not as parts of an interconnected whole. While this is true in many instances, there also have existed for decades numerous, powerful integrating tools like social accounting matrixes and computable general equilibrium models. Where they are concerned, it is not a lack of integration that is the stumbling block but inadequate or inaccurate specification in the integrating models.

A second outcome is a focus on transitory panaceas. Big data and randomized control trials are contemporary examples. Bettencourt (2013) connects the two in hopes that the combination of "new analyses of large urban data sets" will lead to "(p)olicy . . . experiments, that, if carefully designed and measured, can [real-ize] a new science of performance-based planning." But these tools too depend on proper specification, and they also fall short when not calibrated to capture systemic forces.

Transient panaceas abound on the policy side too, from the Garden Cities movement at the turn of the 20th century to the New Town and Satellite City enthusiasms of mid-20th century, the lure of Smart Cities and New Urbanisms at end-century, and already, so early in this one, to the rise (2002) and rapid decline (2016) of the "Creative Class" rendition of the urban cure (Florida 2002, 2016). Author Florida in 2002 touted the leading role of educated and wealthy workers in "the creative professions" in sustaining healthy cities, and recommended that municipalities prioritize attracting this class of people over companies with cool bars, shabby-chic coffee shops and art venues in order to "magically restore our cities." But he, and the cities that adopted his formula, did not contend with the "very deep, dark side" of the urban creative revolution that "bifurcated commu-nities through gentrification, unaffordability, segregation, and inequality" while providing economic growth for the already rich. These forces and outcomes are the focus of his *mea culpa*.

A third upshot is the regular call for more and better information. This too is welcome, but it is not a paucity of information that is the lacuna here. Generations of scholars and practitioners have responded to earlier summons for more and bet-ter information with innumerable studies from every part of the globe on almost every conceivable urban issue. The data we have already are plentiful but remain inadequate because of a missing systemic element. It is the issues not raised and the questions not asked that are at the crux here, not the need to collect more infor-mation for long-established lines of inquiry.

A fourth pitfall is urban goal setting by comparison. A prominent example is the preoccupation in some Indian and international quarters with having Mumbai compete with Shanghai. Former Prime Minister Manmohan Singh stated in 2004 that:

> (W)hen we talk of a resurgent Asia, people think of the great changes that have come about in Shanghai. I share this aspiration . . . to transform Mumbai in the next five years in such a manner that people would forget about Shang-hai and Mumbai will become a talking point.
> (Speech in Mumbai October 6, 2004, quoted in Rediff India Abroad)

and a 2003 McKinsey Report had proposed an "eight-pronged program" to achieve just that (McKinsey and Company Ltd. 2003). Now Mumbai's problems are long known and well documented, as are the steps needed to remedy them. The key is to make health, education, and economic opportunity a priority for all citizens and not just a privileged few. Clearly this is much harder to achieve than to prescribe, but just as clearly the cities that have made the desired transition have done so on the strength of internal commitments to self-betterment and much less or not at all from competitive aspirations.

This is the proverbial monster in the analytical room. Leo Tolstoy captured the situation perfectly when he described his relationship with his serfs:

> The present position which we, the educated and well-to-do classes, occupy is . . . riding on a poor man's back . . . we are sorry for the poor man, very sorry. And we will do almost anything for the poor man's relief; we will not only supply him with food sufficient for him to keep on his legs, but we will provide him with cooling draughts concocted on strictly scientific principles; we will teach and instruct him . . . and give him lots of good advice. *Yes, we will do almost anything but get off his back* (emphasis added).
>
> (Addams *et al.* 1893)

When systemic elements do get factored into urban analyses, urbanization is understood as a dependent as well as independent variable. Cities chart their own courses but their autonomy is proscribed by the goals and mores of a historical phase of development. We saw with Gordon, Hymer, and Harvey that the generic location determining economies of scale and agglomeration are subject to the particular goals and mores of their host societies.[18] Urbanization alone neither fully causes nor fully explains development.

Yet, urbanization continues to be conflated with development worldwide. China made it a key goal two decades ago in the belief that cities "drive not only consumption and investment demand, and create employment opportunities, but directly affect the well-being of people" (Premier Li Kequiang quoted in MacMahon 2013). The strategy has had mixed results as:

> China's gleaming ghost cities have drawn neither jobs nor people . . . while housing shortages have approximately 130 million Chinese migrants living in old shipping containers and tiny, subdivided rooms rented out by the former farmers whose villages have been surrounded by sprawl.
>
> (Hornby & Lee 2013)

This seems not to have dissuaded the country's State Council, which announced a fresh initiative in March 2014 to shift another hundred million people into cities in the next six years, still calculating that urbanization "is a strong engine to keep economic growth . . . on a healthy track" (Johnson 2014).

Excessive belief in the primordial powers of urbanization is not limited to China or to poorer nations. Both leading parties in India's 2014 national election vied to "fetch urban areas the moon" (Urban Update 2014) with competing promises to create a hundred new urban centers. India's Prime Minister Narendra Modi is on record emphasizing urbanization's essential role as a developmental "lifeline for the poor." Major US scholars also tout urbanization as a primary developmental force (Hollis 2013; Glaeser 2011; Brook 2013). Vishaan Chakrabarti would have the federal government make the nation "a more prosperous, sustainable and equitable country" by favoring cities over suburbs to create "a more urban America" (Chakrabarti 2014). Robert Stern also promotes an urban "planned paradise," but intriguingly in reverse, by investing in as opposed to neglecting suburbs, to counter the "tragically interrupted 150-year-old tradition" of Garden Suburbs (quoted in Arieff 2014).

Proponents of urbanization for itself tend to endow it with powers that are not its proper preserve. Cities compared to rural areas have more of the amenities and characteristics associated with development and urbanization gets the credit. Conversely, the problems of cities, especially large ones, are something to "exponential urbanization." But urbanization in both cases is not so much a cause as much as a symptom and agent of the underlying social forces whose effects range far beyond towns and cities to include the entire planet (Eisenstadt & Schlar 1987). Because it presently is in runaway mode its problems are more visible, but cities, even mega ones, are no more the cause of the imbroglio than disappearing glaciers are the cause of rising seas. Both are outward manifestations of other and deeper forces that need to be unearthed and diagnosed; in one case, what causes the glaciers to melt in the first place, in the other, why cities form and operate as they do.

Cities without capital

We addressed that question of cities in capital by tracing the urban impact of the system's tendencies of movement. Three things became clear: that the tendencies generate concurrently both plenty and want, that the former is not achievable without the second, and that system imperatives detract from basic needs reaching all inhabitants of a capitalist society. We noted also the key importance of systemic analyses in urban issues.

A template for future cities without capital therefore would seek to retain the system's productive strengths while shedding the human and environmental costs that have come with it.

This was attempted by the handful of countries that opted out of capital in the last century. The topic is too vast, and this author unqualified, to deal with it except in one small instance. As a native Indian working in economic development, I kept up semi-professionally with changes in the People's Republic of China, particularly in its formative first decade. First-hand accounts by US scholars and reporters who were on the scene in the late 1940s in Shanghai (Schuman 1979), Beijing (Bodde 1950), and Canton (Biggerstaff 1979), plus early histories of Tientsin (Lieberthal 1980), and Barnett (1964) attest to what can be achieved for improving human well-being even in very poor countries when decisions are made for the direct attenuation of human needs unburdened by the profit motive.

I witnessed some of the long-term benefits of these initiatives on my first visit to Beijing and Shanghai in 1985. One of my colleagues on the trip was architect and planner Ed Bacon. Ed had worked in Shanghai 50 plus years ago, in 1933–34, and traveled across China. Olpadwala 2012 recounts our reactions to the two Chinese "cities without capital" (at that time). Ed's is a before and after appreciation, mine a comparison with very similar cities in capitalist India.

Ed was struck by the vast improvements in sanitation, health, nutrition, housing, and public transportation. He remarked on the absence of any visible poverty and how green Beijing had grown. I too had noted these features, in contrast to the state of affairs in Delhi and Calcutta. Both of us were impressed by the non-servile attitude of service personnel (wait staff, cleaners, drivers) compared to old China and contemporary India.

There are offsetting costs to this progress that the article highlights but China's experience suggests that cities without capital can make a positive difference in the daily lives of vulnerable populations. Hopefully future transitions will learn to improve on the process.[19]

Table 1.1 Business motivation, behaviors, and expected outcomes

Business motivation	Expected outcomes/tendencies
1. Private production	a) dissonance between private and public cost calculus
	b) Inequality from catering to effective demand
	c) externalization of costs
	d) concentration
2. Competition	*move towards lowest common denominator
3. Unlimited financial accumulation	a) no inbuilt limit to growth
	b) concentration
Business behaviors	
Increase revenues	
4. Increase market share	
locally	a) growth, concentration, employment, crises
abroad	b) growth, concentration, employment, internationalization, crises
5. Create new products	*growth, concentration, technological change, commoditization
6. Increase selling prices	*growth, concentration, crises
7. Speed the turnover capital	*technological change
8. Maximize public support (subsidies, grants)	*concentration
Reduce costs	
9. Decrease quantity of inputs	
Material	a) technological change
Human	b) technological change, unemployment, inequality, crises
10. Decrease inputs prices	
Increase market power	a) concentration, inequality, crises
Buy globally	b) internationalization, unemployment, crises
11. Minimize public obligations (taxes, regulations)	*concentration

Table 1.2 Business tendencies

1	Growth
2	Commoditization
3	Concentration
4	Technological dynamism
5	Internationalization
6	Unemployment
7	Inequality
8	Crises

Notes

1 For growth, see Diamandis and Kotler 2012; Gordon 2016; for commoditization, Sandel 2012; for concentration, Grullon, Larkin & Roni Michaely 2016; Galston and Hendrickson 2018; Doidge, Karolyi & Stulz 2017; for technological dynamism, Landes 1969; Bernal 1971; Noble 1979, 1986; Winner 1978; for internationalization, Braudel 1982–84; Dalrymple 2019; de Grazia 2005; Findlay and O'Rourke 2007; for unemployment U.S. Bureau of Labor Statistics and OECD publications; for inequality, Piketty 1997, 2014; Bartels 2008; Milanovic 2010; Bourguignon 2012; Stiglitz 2012; and for crises, NBER 2019; Reinhart & Rogoff 2009; Shaikh & Tonak 1994; Shaikh 2016.

2 Early schemes went by weight and perishability of natural resources (von Thunen). Later production played a larger part (Christaller, Losch). Weber brings in agglomeration. Latter three schemes abstract with many simplifying assumptions.

3 Arab scholars.

4 Stull pieces together Smith's urban hierarchy consisting of four levels from a host of scattered references in *The Wealth of Nations* (Table 2, page 303 in Stull). The capital, London, occupies the highest level, followed by "great towns" (Birmingham, Edinburgh, Manchester, etc.), "small market towns and country villages," and "very small villages." The hierarchy rests on the progressive division of labor, with productivity, per capita income, quality of goods made, trade, population, and the urban share of population "increasing together in a mutually reinforcing upward spiral" (p. 304).

5 Hymer (1979b:157) and Harvey (1985:195) echoed this some years later: "Urban centers can then become centers of coordination, decision-making, and control, usually within a hierarchically organized geographical structure."

6 Many others have highlighted the deep link between capital and urbanization, for example, economists Malthus, Smith, as we have seen, Marx, Dobb (1963), and Sweezy, plus historians Pirenne, Braudel, and Hobsbawm, but Harvey posits a deeper, visceral connection.

7 Harvey leaves out the crucial fifth element of obfuscating the transfer of surplus from producers to owners of capital. This fifth factor is central to capital's fundamental canard that it is the fairest of economic systems and the only one that guarantees all participants the full value of the marginal product of their labor.

8 Hymer (1971). The author uses the story of Robinson Crusoe as an allegory for the primitive accumulation phase of capitalist development. The quote expresses the standard reading of Crusoe personality and achievements and Hymer goes to show that the actual story is not of one who "masters nature through reason" but "through conquest, slavery, robbery, conquest and force."

9 Beauregard covers a litany of services from physical (roads, trains, airports, wharves, sanitation) to social (health, education, banking) to legal (police, zoning, protection of private property, interest rate setting).

10 Harvey's survey covers the post-World War II infrastructure boom in the United States with suburbanization, metropolitan re-engineering, and interstate highways, "roller coaster crises" in East and South East Asia in 1997–98, Russia in 1998, and Argentina in 2001, property market booms in Britain, Chile, and Spain, and, especially, vast debt-financed infrastructure projects in China, plus "criminally absurd mega-urbanization projects in such places as Dubai and Abu Dhabi."

11 The apparent discrepancy between a shortfall of one or more million units against a need of "only" a half million homeless may be reconciled by noting that that the two numbers, while clearly linked in reality, are estimated separately for separate purposes. FreddieMac's (2018a) estimates are the difference between housing unit production and the total of household growth, replacement need of deteriorated units, a provision for second homes (!) and a factor added for market solvency. This figure includes also people who are not homeless but who, not having the wherewithal to buy homes, stay on in "second best" situations, for example, young adults with their parents. The homeless estimate is an average number for any given night, with the total number higher than a half million.

12 The situation is not limited to New York. "Parts of London have become . . . more expensive and more empty. At some times of the year the areas are virtually abandoned. An American who has lived in Belgrave Place said that the quiet could become oppressive" (Lyall 2013). Blackstone and other large Wall Street firms took advantage of the mortgage meltdown a decade ago to buy "foreclosed homes by the thousand from courthouse steps, often sight unseen" (Dezember 2017).

13 A half-century of the War on Poverty entailed an expenditure of $15 trillion (Tanner 2012).

14 A recent novel twist is the *micro level commoditization* of private residences through initiatives like Airbnb.

15 "They toted duffels stuffed with millions of dollars in cashier's checks made out in various denominations so they wouldn't have to interrupt their buying sprees with trips to the bank" (Dezember and Kussito 2017).

16 This is true even at the best of times. Infant mortality in Sao Paulo, Brazil's richest city and the heart of the Brazilian Miracle of the 1960s, rose 45 percent between 1960 and 1973 at a time when the national economy was growing at 10 percent a year (Sao Paulo Peace and Justice Commission 1978).

17 Marcuse 2012 is an excellent example. The author uses systemic elements, starting with commoditization, to analyze the state of housing and formulate well-grounded policies.

18 Olpadwala 2000 discusses the political economy of the built environment in detail.

19 Early post-revolutionary experience was similar in Soviet and Cuban cities. Historian Robert Service, no apologist for the Soviet Union, remarks that "Bolsheviks . . . aimed to provide everyone with an abundance of material and cultural well-being. Schooling and health care were already free of charge. Wherever possible, housing was made available for the poor. . . . They were the most ambitious modernizers the country had known since Peter the Great" (Service 2012:330). And I.F. Stone has observed: "It is not because of the undoubted evils in Russia that war is being mustered against her; the same men are only too ready to compromise with the same evils elsewhere and to use the same evil methods at home. It is the good, not the evil, in the Soviet Union that stirs their hate" (Stone 1951).

References

Addams, J. *et al.*, 1893, *Philanthropy and social progress seven essays*, pp. 201–202, Norwood Press, Boston, MA.

Arieff, A., 2014, 'Can paradise be planned?', *New York Times*, 18 April.

Barnett, A.D., 1964, *Communist China: The early years 1949–55*, Pall Mall Press, London.

Bartels, L., 2008, *The new gilded age*, Princeton University Press, Princeton, NJ.

Beauregard, R.A., 2018, *Cities in the urban age: A dissent*, The University of Chicago Press, Chicago, IL.

Bernal, J.D., 1971, *Science in history*, MIT Press, Cambridge, MA.

Bettencourt, L.M.A., 2013, 'The origins of scaling in cities', *Science* 340(6139), 1438–1441.

Biggerstaff, K., 1979, *Nanking letters*, Cornell University East Asia Papers, Ithaca, NY.

Bodde, D., 1950, *Peking diary: A year of revolution*, Henry Schuman, Inc., New York, NY.

Bourguignon, F., 2012, *The globalization of inequality*, Princeton University Press, Princeton, NJ.

Braudel, F., 1982–84, *Civilization and capitalism*, Harper and Row, New York, NY.

Brook, D., 2013, *History of future cities*, Norton, New York, NY.

Chakrabarti, V., 2014, 'America's urban future', *New York Times*, 17 April.

Christaller, W., 1966, *Central places in Southern Germany*, Prentice-Hall, Englewood Cliffs NJ.

Coalition for the Homeless, 2020, www.coalitionforthehomeless.org/wp-content/uploads/2020/02/NYCHomelessShelterPopulation_Worksheet_1983-Present_Dec2019.pdf.

Cohen, E., 2019, 'Ony Washington can solve the nation's housing crisis', *New York Times*, 10 July.

Dalrymple, W., 2019, *Anarchy: The East India company, corporate violence, and the pillage of an empire*, Bloomsbury, London.

de Grazia, V., 2005, *Irresistible empire: America's advance through 20th century Europe*, Belknap, Harvard University Press, Cambridge, MA.

Dewan, S., 2014, 'In many cities, rent is rising out of reach of middle class', *New York Times*, 15 April.

Dezember, R., 2017, 'Blackstone, starwood to merge rental home businesses in bet to be America's biggest home landlord', *Wall Street Journal*, 10 August.

Dezember, R. & Kussito, L., 2017, 'Meet your new landlord: Wall Street', *Wall Street Journal*, 21 July.

Diamandis, P. & Kotler, S., 2012, *Abundance: The future is better than you think*, Free Press, New York, NY.

Dobb, M., 1963, *Studies in the development of capitalism*, International Publishers, New York, NY.

Doidge, C., Karolyi, G.A. & Stulz, R.M., 2017, 'The U.S. left behind? Financial globalization and the rise of IPOs outside the U.S.', *Journal of Financial Economics* 123, 3 March.

Editorial Board, 2020, 'Rentals everywhere, but no place to live: Government rules drive developers to build luxury apartments', *Wall Street Journal*, 5 February, https://www.wsj.com/articles/rentals-everywhere-but-no-place-to-live-11580948368.

Einstein, A., 1949, 'Why socialism', *Monthly Review* 1(1).

Eisenstadt, S.N. & Schlar, A., 1987, *Culture and urbanization*, Sage Publications, Newbury Park, CA.

Eviction Lab Updates, 2018, 'National estimates: Eviction in America', in *The eviction lab*, Princeton University, Princeton, NJ, 11 May.

Findlay, R. & O'Rourke, K.H., 2007, *Power and plenty: Trade, war and the world economy in the second millennium*, Princeton University Press, Princeton, NJ.

Florida, R., 2002, *The rise of the creative class*, Basic Books, New York, NY.

Florida, R., 2016, *The new urban crisis*, Basic Books, New York, NY.

FreddieMac Insight, 2018a, *Locked out? Are rising costs barring young adults from buying their first homes?*, 28 June, www.freddiemac.com/research/insight/20180628_rising_housing_costs.page.

FreddieMac Insight, 2018b, *The major challenge of inadequate housing supply*, 5 December, www.freddiemac.com/research/insight/20181205_major_challenge_to_u.s._housing_supply.page#ResearchChart1.

Galston, W.A. & Hendrickson, C., 2018, *A policy at peace with itself: Antitrust remedies for our concentrated, uncompetitive economy*, The Brookings Institution, Washington, DC, 5 January, viewed 19 September 2018, from www.brookings.edu/research/a-policy-at-peace-with-itself-antitrust-remedies-for-our-concentrated-uncompetitive-economy/.

Glaeser, E., 2011, *The triumph of the city: How our greatest invention makes us richer*, Penguin, New York, NY.

Gordon, D., 1976, 'Capitalist development and the history of American cities', in W.K. Tabb & L. Sawyers (eds.), *Marxism and the metropolis*, Oxford University Press, Oxford.

Gordon, R., 2016, *The rise and fall of American growth: The U.S. standard of living since the civil war*, Princeton University Press, Princeton, NJ.

Grullon, G., Larkin, Y. & Michaely, R., 2016, *Are US industries becoming more concentrated?*, October, viewed 19 September 2018, from https://finance.eller.arizona.edu/sites/finance/files/grullon_11.4.16.pdf.

Harvey, D., 1985, *The urbanization of capital*, The Johns Hopkins University Press, Baltimore, MD.

Harvey, D., 2008, 'The right to the city', *New Left Review* 53, September–October.

Holder, S., 2018, 'Where commuting is out of control', *Citylab*, 25 April, www.citylab.com/transportation/2018/04/where-commuting-is-the-worst/558671/.

Hollis, L., 2013, *Cities are good for you: The genius of the metropolis*, Bloomsbury Publishers, New York, NY.

Honan, K., 2018, 'New York city's affordable housing units dwindled since 2005', *Wall Street Journal*, 25 September.

Hornby, L. & Lee, J.L., 2013, 'In China's urbanization, worries of a housing shortage', *New York Times*, 31 March.

HUD, 2018, *The 2018 annual homeless assessment report (AHAR) to congress*, Department of Housing and Urban Development, Washington, DC.

Hymer, S., 1971, 'Robinson Crusoe and the secret of primitive accumulation', *Monthly Review* 23(4), September.

Hymer, S., 1979a, 'The multinational corporation and the law of uneven development', in R.B. Cohen *et al.* (eds.), *The multinational corporation: A radical approach: Papers by Stephen Hymer*, Cambridge University Press, Cambridge.

Hymer, S., 1979b, 'The multinational corporation in the international division of labor', in R.B. Cohen *et al.* (eds.), *The multinational corporation: A radical approach: Papers by Stephen Hymer*, Cambridge University Press, Cambridge.

Johnson, I., 2014, 'China approves plan to increase urban population: Aims to have about 60% of population living in cities by 2020', *Wall Street Journal*, 18 March.

Joint Center for Housing Studies of Harvard University, 2019, *The state of the nation's housing 2019*, Cambridge, MA, www.jchs.harvard.edu/state-nations-housing-2019.

Joint Center for Housing Studies of Harvard University, 2020, *America's rental housing 2020*, Cambridge, MA, www.jchs.harvard.edu/americas-rental-housing-2020.

Kingsley, G.T., 2017, *Trends in housing problems and federal housing assistance*, Metropolitan Housing and Communities Policy Center, Urban Institute, Washington, DC, www.urban.org/sites/default/files/publication/94146/trends-in-housing-problems-and-federal-housing-assistance.pdf.

Krugman, P., 2017, 'Why can't we get cities right?', *The New York Times*, 4 September.

Landes, D.S., 1969, *The unbound prometheus technological change and industrial development in Western Europe from 1750 to the present*, Cambridge University Press, Cambridge.

Lieberthal, K.G., 1980, *Revolution and tradition in Tientsin 1949–52*, Stanford University Press, Stanford, CA.

Losch, A., 1954, *The economics of location*, Yale University Press, New Haven, CT.

Loudenback, T., 2019, 'The U.S. has more million dollar markets than ever – here are 23 cities where the typical home will be $1 million by next year', *Business Insider*, 6 November.

Lyall, S., 2013, 'A slice of London so exclusive even the owners are visitors', *The New York Times*, 1 April.

MacMahon, D., 2013, 'China's gleaming ghost cities draw neither jobs nor people', *Wall Street Journal*, 9 August.

Marcuse, P., 2011, 'A critical approach to solving the housing problem', in N. Brenner, P. Marcuse & M. Mayer (eds.), *Cities for people, not for profit*, Routledge, New York, NY.

McKinsey and Company Ltd., 2003, *Vision Mumbai: Transforming Mumbai into a world class city*, McKinsey and Company Ltd., Mumbai and New Delhi, September.

Milanovic, B., 2010, *The haves and have nots: A brief and idiosyncratic history of global inequality*, Basic Books, New York, NY.

National Alliance to End Homelessness, 2019, *State of homelessness report*, National Alliance to End Homelessness, Washington, DC.

National Association of Home Builders (NAHB), 2018, *Nation's stock of second homes*, 6 December, http://eyeonhousing.org/2018/12/nations-stock-of-second-homes/.

National Bureau of Economic Research, *U.S. business cycle expansions and contractions*, Washington, DC, viewed 12 June 2019, from www.nber.org/cycles.html.

National Center for Homeless Education, 2020, *Federal data summary school years 2015–16 through 2017–18: Education for homeless children and youth*, University of North Carolina, Greensboro, NC.

Neuman, W., 2018, 'As 4 out of 5 in public housing lost heat, a demand for an apology is unfulfilled', *New York Times*, 6 February.

Noble, D., 1979, *America by design: Science, technology and the rise of corporate capitalism*, Oxford University Press, New York, NY.

Noble, D., 1986, *Forces of production: A social history of industrial automation*, Oxford University Press, New York, NY.

Olpadwala, P., 2000, 'The political environment of the built environment', in P. Knox & P. Ozolins (eds.), *Design professionals and the built environment: An introduction*, John Wiley and Sons, Ltd., West Sussex.

Olpadwala, P., 2012, 'Cities and classes in early-reform China and India', in K.-M. Liu (ed.), *My first trip to China: Scholars, diplomats and journalists reflect on their first encounter with China*, East Slope Publishing, Hong Kong.

Olpadwala, P., 2019, *Commonsense Marx: Deducing capital's tendencies*, ResearchGate.

Olpadwala, P. & Mansury, Y., 2007, 'Finance and production in the United States, 1928–2001: An empirical note', in A.K. Bagchi & G.A. Dymski (eds.), *Capture and exclude: Developing economies and the poor in global finance*, Tulika Books, New Delhi.

Piketty, T., 1997, *The economics of inequality*, Harvard University Press, Cambridge, MA.

Piketty, T., 2014, *Capital in the twenty first century*, Belknap, Harvard University Press, Cambridge, MA.

Rediff India Abroad, www.rediff.com.news/2004/oct/06pm.htm.

Reinhart, C.M. & Rogoff, K.S., 2009, *This time is different: Eight: Centuries of financial folly*, Princeton University Press, Princeton, NJ.

Sandel, M.J., 2012, *What money can't buy: The moral limits of markets*, Farrar, Straus and Giroux, New York, NY.

Sao Paulo Peace and Justice Commission, 1978, *Sao Paulo: Growth and poverty*, Sao Paulo Peace and Justice Commission, Sao Paulo.

Schuman, J., 1979, *China: An uncensored look*, Second Chance Press, Sagaponack, NY.

Service, R., 2012, *Spies and commisars: The early years of the Russian revolution*, Public Affairs, New York, NY.

Shaikh, A., 2016, *Capitalism: Competition, conflict, crises*, Oxford University Press, New York, NY.

Shaikh, A. & Tonak, E.A., 1994, *Measuring the wealth of nations*, Cambridge University Press, Cambridge.

Sharf, S., 2017, 'For sale: 18 homes seeking $100 million or more', *Forbes*, 10 August.

Shen, L., 2016, 'Here's how many homes the average billionaire now owns', *Fortune*, 2 December.

Solomon, E.B., 2019, 'NYC's ghost towers', *The Real Deal*, 1 April, https://therealdeal.com/issues_articles/ghost-towers-new-york-city/.

Stiglitz, J.E., 2012, *The price of inequality: How today's divided society endangers our future*, W.W. Norton, New York, NY.

Stone, I.F., 1951, *The daily compass*, 16 March.

Stull, W.J., 1986, 'The urban economics of Adam Smith', *Journal of Urban Economics* 20(X).

Tanner, M., 2012,'The American welfare state: How we spend nearly $1 trillion a year fighting poverty – and fail', *Policy Analysis* 694, 11 April.

Thompson, D., 2020, 'Why Manhattan's skyscrapers are empty', *The Atlantic*, 16 January.

Timiraos, N., 2015, 'Has industry consolidation held back construction?', *Wall Street Journal*, 8 September.

UrbanUpdate, 2014, *Congress, BJP vie to fetch urban India the moon*, 8 April, http://urbanupdate.in/infrastructure/congress-bjp-vieurban-india-moon.

Urban Wire, 2020, 'Our aging public housing puts older Americans at risk', in *Urban wire*, Urban Institute, Washington, DC.

USA Today, 2013, 'Sky's the limit: New towers for the rich soar in New York', *USA Today*, 14 December.

U.S. Census Bureau, 2020a, 'Housing units July 1, 2018', *Quick Facts*, www.census.gov/quickfacts/fact/table/US/VET605218.

U.S. Census Bureau, 2020b, 'Vacancy Rates for the U.S.', *Housing Vacancies and Homeownership (CPS/HVS): 2019*, www.census.gov/housing/hvs/data/q419ind.html.

Velazquez Media Center, 2019, *Representative Nydia M. Velazquez seeks funding for public housing*, https://velazquez.house.gov/media-center/press-releases/vel-zquez-seeks-funding-public-housing.

Von Thunen, J.H., 1966, *'Isolated state'*, Pergamon, New York, NY.
Weber, A., 1929, *Theory of location of industry*, University of Chicago Press, Chicago, IL.
White House Data, table B-2, p. 324 and B-17, p. 345, in Appendix B, www.whitehouse.
gov/sites/default/files/docs/erp2013/ERP2013_appendix_B.pdf.
Winner, L., 1978, *Autonomous technology: Technics out-of-control as a theme in political thought*, MIT Press, Cambridge, MA.

2 Cities and subjectivity within and against capitalism

Kanishka Goonewardena and Sinead Petrasek

City/country/contradiction

'City air liberates' (*Stadtluft macht frei*)! This popular slogan referring to a provision of early medieval Germanic law, which offered a novel measure of urban freedom to serfs fleeing from their lords, offers a classic illustration of the association of the modern city with the promise of human freedom and flourishing – of the real and imagined historic correlation of urban society with political emancipation and material prosperity – though never unequivocally. In *Caliban and the Witch*, Silvia Federici notes that 'by the 13th century', women 'were leading the movement away from the country, being the most numerous among the rural immigrants to towns', forming by the 15th century 'a large percentage of the population of cities' in Western Europe. 'City laws did not free women' and 'few could afford to buy the "city freedom", as privileges connected with the city life were called'. Yet, 'in the city, women's subordination to male tutelage was reduced, as they could now live alone'. Even though 'most of them lived in poor conditions', according to Federici, urban life 'gave them a new social autonomy'.[1]

It was acknowledged – in Henri Pirenne's *Medieval Cities* (1925) or Max Weber's *The City* (1927)[2] – that life in the medieval European city was not so easy for runaway serfs either, who often ended up with meagre means of survival as servants or slaves – proletarians in the classical sense, anticipating the struggles of their modern progenies.[3] In *Society of the Spectacle* (1967) Guy Debord too invokes the popular appeal of liberating city air. 'But', he adds, 'while the history of cities is certainly a history of freedom, it is also a history of tyranny'.[4] Still, the desire to link cities with the good life has been – though not unqualified – durable and powerful.

The long-standing promise of the city – as much as the 'idiocy of rural life' notoriously noted in *The Communist Manifesto* – begs therefore to be dialectically understood. Our methodological role models here may well be Marx and Engels, who grasped the historical dynamics of capitalist development as inherently contradictory – at once emancipatory and alienating. As Marshall Berman underlines in *All That Is Solid Melts Into Air* (1982),[5] Marx and Engels appreciated with unparalleled force the revolutionary accomplishments of the bourgeoisie, while demonstrating in equally powerful terms the injustice and unsustainability of the capitalist way of life, which must in turn be overcome by the democratically assembled powers of working people. Similar attention to the contradictions of cities – and the dialectical method they demand – is warranted by the constitutive relationship between urbanization and capitalism, which was prophetically glimpsed by the 24-year-old Engels in his classic study, *The Condition of the Working Class in England* (1993/1845).[6]

Approaching London along the Thames, Engels finds the city 'so vast, so impressive, that a man cannot collect himself, but is lost in the marvel of England's greatness before he sets foot upon English soil'. He seems mesmerized by '[t]his colossal centralization, this heaping together of two and a half millions of human beings at one point' that is London, which 'has multiplied the power of this two and a half million a hundredfold' – as if anticipating Henri Lefebvre's theorization of 'centrality' as a supreme formal feature of urban life.[7] The wonders of this urban agglomeration economy and density of social life appear in a more complex light, however, '[a]fter roaming the streets of the capital a day or two'. Now it looks like 'these Londoners have been forced to sacrifice the best qualities of their human nature, [in order] to bring to pass all the marvels of civilization'. For Engels, '[t]he very turmoil of the streets has something repulsive, something against which human nature rebels'.[8] The intensely contradictory form of this centrality is revealed in the nature of the mass of people in the city – the crowd. It is at once social and individual, and rendered well-nigh incapable of reconciling these fundamental but opposed aspects of the modern metropolis.

People 'crowd by one another as though they had nothing in common, nothing to do with one another, and their only agreement is a tacit one, that each keep to his own side of the pavement, so as not to delay the opposing streams of the crowd, while it occurs to no man to honour another with so much as a glance'.[9] Here Engels registers with an ethnographer's eye, not only how 'the great towns' bring people together in unprecedented power and potential, but also how they pull them apart:

> The brutal indifference, the unfeeling isolation of each in his private interest becomes the more repellent and offensive, the more these individuals are crowded together, within a limited space. And, however much one may be aware that this isolation of the individual, this narrow self-seeking is the fundamental principle of our society everywhere, it is nowhere so shamelessly barefaced, so self-conscious as just here in the crowding of the great city. The dissolution of mankind into monads, . . . the world of atoms, is here carried out to its extreme.[10]

Engels is among the first to theorize the relationship of such urbanization to modes of production, underlying the inherent contradictions of both. In the contemporaneous but posthumously published work *The German Ideology* (1845), Marx and Engels write: 'The division of labour inside a nation leads at first to the separation of industrial and commercial from agricultural labour, and hence to the separation of *town* and *country* and to the conflict of their interests'.[11] Not only are the town and country themselves internally contradictory, but so is the historical relationship between them, by virtue of the hierarchical division of social labour and underlying forms of property relations: 'tribal property', 'ancient communal and state property', 'feudal or estate property' and 'modern private property' characteristic of capitalism.

> The most important division of material and mental labour is the separation of town and country. The contradiction between town and country begins with the transition from barbarism to civilisation, from tribe to state, from locality to nation, and runs through the whole history of civilization to the present day.[12]

'The contradiction between town and country', for Marx and Engels, 'can only exist with the framework of private property'. As such, '[t]he abolition of the contradiction between town and country is one of the first conditions of communal life',[13] or what we may call the 'commons'. These historical-geographical reflections in *The German Ideology* become even more pronounced in the *Manifesto*, where Marx and Engels sketch – even if in Eurocentric language – the elements of a radical conception of capitalism's uneven and combined development:

> The bourgeoisie has subjected the country to the rule of the towns. It has created enormous cities, has greatly increased the urban population as compared with the rural, and has thus rescued a considerable part of the population from the idiocy of rural life. Just as it has made the country dependent on the towns, so it has made barbarian and semi-barbarian countries dependent on the civilised ones, nations of peasants on nations of bourgeois, the East on the West.[14]

Hence the ninth of the ten political measures advocated in the *Manifesto*'s programmatic second section entitled 'Proletarians and Communists' – namely, the 'gradual abolition of the distinction between town and country', which implies at once the abolition of the unequal social division of labour characteristic of capitalist production and reproduction.[15] Marx and Engels's historic conception of communism includes a spatial as much as a social dimension, demanding a transcendence of town, country and the relationship between the two as we know them.

So urbanization is as fraught as capitalism, with its promising and alienating moments inextricably intertwined. The form of this contradiction recalls the very nature of modernity, at once the prospect for liberation from oppressive tradition and the threat of nihilist destruction – of social as much as natural life. In his celebrated work on modernism, Berman writes that '[t]o be modern is to live a life or paradox and contradiction'.[16] For students of urbanization, this conception of modernity is instructive:

> There is a mode of vital experience – experience of space and time, of self and others, of life's possibilities and perils – that is shared by men and women all over the world today. I will call this body of experience 'modernity'. To

be modern is to find ourselves in an environment that promises us adventure, power, joy, growth, transformation of ourselves and the world – and, at the same time, that threatens to destroy everything we have, everything we know, everything we are . . . To be modern is to be part of a universe in which, as Marx said, 'all that is solid melts into air'.[17]

Being properly modernist, then, means grasping the contradictions of modernity radically, eschewing the temptation of one-sided and un-dialectical perspectives on a world dominated by capitalism. Berman therefore urges us to follow 'nineteenth-century thinkers [who] were simultaneously enthusiasts and enemies of modern life', not so much their 'twentieth century successors [who] have lurched far more towards rigid polarities and flat totalizations'[18]: more Baudelaire, Nietzsche and Marx; less positivist modernization theory and post-colonial self-orientalization. His interpretive horizon is not limited by 'either/or' standpoints, but expanded by 'both/and' dialectics.[19] With regard to our cultural-ideological understanding of the city and country, rarely does Berman's prescription appear more persuasively than in Raymond Williams's *The Country and the City* (1973):

On the country has gathered the idea of a natural way of life: of peace, innocence, and simple virtue. On the city has gathered the idea of an achieved centre: of learning, communication, light. Powerful hostile associations have also developed: on the city as a place of noise, worldliness and ambition; on the country as a place of backwardness, ignorance, limitation.[20]

At a time when the so-called 'urban age' is universally celebrated in the most uncritical fashion – and populist-technotopic appeals to the 'creative class' and 'smart cities' masquerade as solutions to the most pressingly systemic social and ecological crises – the political value of the dialectical method exemplified by Berman and Williams cannot be overstated. In this regard, Fredric Jameson's pithy appreciation of dialectics – and Marx's use of it – is apt:

Marx powerfully urges us to do the impossible, namely, to think . . . [historical] development positively and negatively all at once; to achieve, in other words, a type of thinking that would be capable of grasping the demonstrably baleful features of capitalism along with its extraordinary and liberating dynamism simultaneously within a single thought, and without attenuating any of the force of either judgement. We are somehow to lift our minds to a point at which it is possible to understand that capitalism is at one and the same time the best thing that has ever happened to the human race, and the worst.[21]

Elsewhere, Jameson usefully reminds us: 'one way in which the dialectic can be defined is as a conceptual coordination of incommensurabilities'.[22] So understood, the dialectical method, which has too often been misunderstood as merely a discourse of binaries, applies naturally to concepts as much as realities addressed by students of space, including 'city' and 'country'. This point has recently been affirmed in their own way by the proponents of the concept of 'planetary urbanization', Neil Brenner and Christian Schmid, as Hillary Angelo notes: although the 19th-century 'metropolis is no longer the characteristic form of urbanism, an understanding of the city and its associated oppositions derived from it remain a dominant interpretive frame'.[23] In this regard, Williams's insight on the categories of 'city' and 'country' – that their 'real history, throughout, has been astonishingly varied' – is still unsurpassed.

> Between the cities of ancient and medieval times and the modern metropolis or conurbation there is a connection of name and in part of function, but nothing like identity. Moreover, in our world, there is a wide range of settlements between the traditional poles of country and city: suburb, dormitory town, shanty town, industrial estate.[24]

It is with due attention to the historical and geographical variations of the notions and realities of city and country – and attendant contestations having to do ultimately with the production and reproduction of socio-spatial relations (class, gender, race) – that we may explore the evolving relations between forces of liberation and domination unleashed by capitalist urbanization. Here we approach this struggle between emancipatory and alienating moments of urban experience – with reference to a few classic radical texts, but also with an eye on some salient contemporary manifestations of urbanism and technology – by way of one leading question: how does urban space mediate ideology and subjectivity?

Space/ideology/subjectivity

The idea of liberating city air (*Stadtluft macht frei*), in its commonsensical form, is a liberal one. Its hegemonic appeal rests on the notion of *individual* liberty, the ultimate promise of modern life in the bourgeois imagination. The force of its attraction is understandable against the feudal backdrop of the ancient regime – or even in oppressive urban conditions today, in spite of proven limitations of libertarian ideology when it comes to actualizing its ideal of freedom. Yet, the dominant conception of 'freedom' as an individual affair also has been eminently questionable; even the great slogan of the French Revolution clearly spelled out a far more radically socialist vision: liberty, equality, fraternity!

The liberal revolutions of Europe relied heavily on the radicalism of the lower estates – workers and peasants – before turning against them and their radically egalitarian ideals of liberty, which could neither be conceived nor actualized without equality and fraternity. More acutely than most thinkers of his time, Hegel – an ardent enthusiast of the French Revolution against Prussian absolutism – demonstrates this in his *Philosophy of Right* (1821), as he registers modern civil society as a contradictory space of both liberty and alienation that needed to be sublated by means of a rationalized state embodying universal reason.[25]

The questions of liberty, equality and fraternity raised by Hegel nearly 200 years ago still remain to be addressed in our own neoliberal time. Their imbrication with the dynamics of capitalist urbanization too has been of paramount if understated importance for socialist-revolutionary politics, as Henri Lefebvre argues in *Le droit à la ville* (1968), *The Urban Revolution* (1970) and *The Production of Space* (1974).[26] In radical urban theory mindful of such matters, a key strategic-theoretical question has centred on the role played by urban space and experience in the making of political subjects within and possibly beyond the contradictions of capitalism – which is a social system cannibalizing structural inequality, along predominantly (but not exclusively) class, gender and race lines, as theorized by revolutionary anti-colonial, anti-racist and socialist-feminist critics.

Prototypes and starting points for such inquiries into the relations between political consciousness and urban experience are present in the writings of Marx and Engels; and perceptive readers of their writings such as Lefebvre, Debord and Walter Benjamin have profoundly enriched left critical-theoretical perspectives on space, ideology and politics. The best of such works, including anti-colonial and socialist-feminist contributions of writers, ranging from Frantz Fanon to Janet Abu-Lughod and Dolores Hayden to Doreen Massey,[27] maintain a dialectical focus on urban space – social, physical and ideological – as a contested terrain, one in which forms of subjectivity both affirmative and critical of the regnant social totality are produced, individual as much as collective.

Marx's well-known book on the revolution and counter-revolution of 1848–1851 in France – *The 18th Brumaire of Louis Bonaparte* (1963/1852)[28] – offers an illuminating reflection on the mediation of ideology by space; or, more specifically, on how under certain socio-spatial conditions the French peasant languished in the notorious 'idiocy of rural life' even as their urban proletarian cousins acquired a revolutionary class consciousness:

> The small-holding peasants form a vast mass, the members of which live in similar conditions but without entering into manifold relations with one another. Their mode of production isolates them from one another instead of bringing them into mutual intercourse. . . . Each individual peasant family is almost self-sufficient; it itself directly produces the major part of its consumption and thus acquires its means of life more through exchange with nature than in intercourse with society. A small holding, a peasant and his family; alongside them another small holding, another peasant and another family. A few score of these make up a village, and a few scores of villages make up a Department. In this way, the great mass of the French nation is formed by simple addition of homologous magnitudes, much as potatoes in a sack of potatoes.[29]

This colourful and controversial passage on the relationship between ideology and space also clarifies what Marx meant by that much-maligned and therefore often misunderstood phrase in the *Manifesto*, 'the idiocy of rural life'. As Eric Hobsbawm points out, although Marx shared with the leading minds of his time the characteristic 'townsman's contempt for – as well as ignorance of – the peasant milieu, the actual and analytically more interesting German phrase ("*dem Idiotismus des Landlebens . . .*") referred not to 'stupidity' but to 'the narrow horizons', or 'the isolation from the wider society', in which the people of the countryside lived. 'Idiocy', in Marx's usage here, closely followed the meaning of its Greek root *idiotes*: 'a person concerned only with his own private affairs and not with those of the wider community'.[30]

An 'idiot', in this sense of the word derived from *idiotes*, is the classical liberal *individual* – from John Locke to neoclassical economics to rational choice theory. But neither Marx nor Engels proposes that such liberal individuals – subjects seeking to maximize their own welfare – are to be found exclusively in the countryside, notwithstanding the substantial support of French peasants for Louis Bonaparte lamented in *The 18th Brumaire*. As noted earlier, Engels saw plenty of them in the London crowd, minding their own business in the streets. But Marx, Engels and other socialists have been also keen students of the potential of cities to nurture radical political consciousness, by bringing people together in their workplaces as well as neighbourhoods. The production of centrality – the dense convergence of social relations and activity – as a formal feature of the urbanization process harbours revolutionary potential for Lefebvre, as he argues in *The Urban Revolution*.[31] Yet, he and others also underline the veritable propensity of capitalist urban space to produce liberal *idiotes*. In this regard, Debord's observations on 'urbanism' are unequivocally dialectical:

> Urbanism is the modern way of tackling the ongoing need to safeguard class power by ensuring the atomization of workers dangerously *massed together* by the conditions of urban production. The unremitting struggle that has had to be waged against the possibility of workers coming together in whatever manner has found a perfect field of action in urbanism . . . But the general trend toward isolation, which is the essential reality of urbanism, must also embody a controlled reintegration of the workers based on the planned needs of production and consumption. Such an integration into the system must recapture isolated individuals as individuals *isolated together*.[32]

Even if 'idiocy' was not confined to rural life but central to urban intercourse as well, the socialist tradition insists that in cities – or for that matter in the countryside, as many revolutions in the colonized world proved in the 20th century – it does not go unchallenged. In fact, *The Condition of the Working Class in England* is above all a product of Engels's learned optimism about the capacity of that class 'in itself' to become a class 'for itself'.

But Engels is also keenly aware of how urban space misrepresents the reality of the city as well as the social totality. In fact, his trenchant remarks on the physical layout of 'the great towns' in England around 1844 provide one of the earliest, and still incomparably lucid, observations on the mediation of ideology by urban space:

> Owing to the curious layout of the town it is quite possible for someone to live for years in Manchester and to travel daily to and from his work without ever seeing a working-class quarter or coming into contact with an artisan. He who visits Manchester simply on business or for pleasure need never see the slums, mainly because working-class districts and the middle-class districts are quite distinct. This division is due partly to deliberate policy and partly to instinctive and tacit agreement between the two social groups. . . . To such an extent has the convenience of the rich been considered in the planning of Manchester that these plutocrats can travel from their houses to their places of business in the center of the town by the shortest routes, which run entirely through working-class districts, without even realizing how close they are to the misery and filth which lie on both sides of the road. . . . [E]ven the less pretentious shops adequately serve their purpose of hiding from the eyes of the wealthy ladies and gentlemen with strong stomachs and weak nerves the misery and squalor which are part and parcel of their own riches and luxury.[33]

Engels's lesson on the role of urban space in representing and misrepresenting reality figures prominently in contemporary urban studies, as exemplified in Mike Davis's iconoclastic study of Los Angeles, *City of Quartz* (1990), as well as in the oeuvres of critics such as Michael Sorkin and Martha Rosler.[34] Theoretically, its implications have been most fruitfully and rigorously developed in Fredric Jameson's work on 'cognitive mapping', where the relationship between space and ideology is examined with special reference to Kevin Lynch's *The Image of the City* (1960) and Louis Althusser's powerful definition of 'ideology' as the 'representation of the subjects' imaginary relationship to their real conditions of existence'.[35]

Jameson's integration of theories of ideology with theories of space – in order to investigate how the production of ideology and the production of space are dialectically related – are resonant with a range of themes investigated in critical theory, especially commodification and reification (Georg Lukács), culture industry (Frankfurt School), hegemony (Antonio Gramsci), discipline and governmentality (Michel Foucault), control (Gilles Deleuze) and subjection (Judith Butler).[36] All of these have significant implications for urban studies, especially in our 'urban age' of creative, smart, competitive and entrepreneurial cities advertising the good life. For several questions raise themselves in the present conjuncture of urbanism, particularly the one about the very meaning and possibility of emancipation promised with the means and ends of neoliberal urbanism.

Reification, subjection, hegemony, control, consumption, governmentality – in their own ways, these conceptions offer vital critiques of the form of liberal individual freedom associated with bourgeois urbanization, exposing the mythic and fetish character of liberty in the face of capital and state logics. As a critique of ideology and discourse – also of attendant social apparatuses and related practices of everyday life – this constellation of concepts reveals as well how the systemic imperatives of capitalism produce subjects calibrated to reproduce rather than sublate the actually existing social totality. With varying degrees of success, these theories also broach the key political question: how could anti-capitalist subjects be produced within capitalism?

Whither cities and subjects?

If we are to imagine a critique of subjectivity and society for the present age that stands in the critical tradition of Marx's theorization of commodification, Max Weber's concepts of disenchantment and rationalization of modernity, Georg Lukács's *History and Class Consciousness* (1923), Theodor Adorno and Max Horkheimer's *The Culture Industry* (1944), Herbert Marcuse's *One Dimensional Man* (1964), or Foucault's more recent ruminations on discipline and governmentality,[37] then it is hard to think of a more pungent and provocative contemporary text than the very short book *Psychopolitics* (2017) by Byung-Chul Han, a Berlin-based critic of Korean origin with Frankfurt School as well as French post-structuralist accents.[38] For few texts pack such power in so few pages devoted to neoliberalism and its latest technologies – with vital implications for the prospects of urban life mediated by big data and social media, amidst a plethora of precarities and anxieties fuelled by financialized capitalism, which in turn feed a fetishism not merely of the commodity, but also of creativity, entrepreneurship and other hyper-curated affectations masquerading as authentic subjectivity.

Foucault, we may recall, wrote in a noted essay that the general theme of his research was not power, but the subject.[39] Here, he argues that the modern liberal state has not so much developed as an entity separate from individual existence, but as one into which individuals are fully and freely integrated, such that the freedom pursued by these individuals conforms to a set of conventions aligned with state objectives. 'It was no longer a question of leading people to their salvation in the next world', writes Foucault, referring to the role undertaken by the modern state, 'but ensuring it in this world'.[40] In this way, the meaning of salvation was transformed from the pastoral sense into a guarantee of material well-being, health and protection under the guardianship of the state. It also allowed for a new form of bio-political management. Thus, for the French philosopher, our political task became no longer the liberation of individuals from the state and its institutions, but something even more challenging and radical – that is, to liberate people from these very forms of individualization that are linked to the state. The way forward, then, would be 'to promote new forms of subjectivity through the refusal of this kind of individuality which has been imposed on us'.[41]

Among critical Left intellectuals, however, Foucault appears to be rather more intrigued by the possibilities of (neo)liberal salvation than most, while also being one of the first to seriously engage with the concepts advanced by (neo)liberal thinkers.[42] He certainly understood (neo)liberalism as a dispensation offering more space for autonomy, plurality and experimentation – in contrast to the disciplinary arrangements of the *ancien regime*. In the process of transforming one's life into an *oeuvre*, this 'self-entrepreneurship' might, Foucault conjectured, result in the 'proliferation of autonomous spaces and the discourse of autonomy'.[43] The (neo)liberal city – and the form of (neo)liberal governmentality that seems to have dazzled Foucault – might take on the role of salvation in the earthly sense.

The optimism was, predictably, a controversial assessment on Foucault's part, as critically noted by Daniel Zamora, who argues that his seductive appeal to living one's life as a work of art was all too uncritical of neoliberalism.[44] Han offers a rather different critical reading of Foucauldian governmentality for the present moment of big data and social media – by dwelling on the subjective dimensions of neoliberalism and technologies of power, while updating Foucault's interest in entrepreneurial 'freedom' to reveal the operations of a more 'totalizing' and novel form of exploitation. According to Han, Foucault failed to grasp how 'the neoliberal regime utterly claims the technology of the self for its own purposes', as it shrewdly compels its subjects to willingly auto-exploit themselves, in addition to continuing with the old-fashioned exploitation of labour that Marx discovered and critiqued.[45] What we can now discern is a kind of *compulsory freedom*, an unending pursuit of individual self-realization that reproduces capitalist relations all the more effectively and pervasively.

For our purposes, what Han points to is a form of entrepreneurial subjectivity engendered by the neoliberal capitalist city. The entrepreneur is no longer the capitalist of yore, defined by ownership of the means of production. It is rather a new kind of subjectivity pertaining to a broader agglomeration of urban subjects engaged in immaterial or creative labour of, who appear to represent the culmination of individual self-actualization and economic freedom, and yet conceal the peril of the so-called 'free' market, as well as the disconnection and atomization felt in everyday life, wherein one's capacity to self-actualize comes at the cost of isolation and exploitation. Referencing Chicago School economist Gary Becker's utility-maximizing agent, writes Han: 'The subject of today's world is an entrepreneur of the self practicing self-exploitation'.[46]

Historically, the entrepreneur (derived from the French *entreprendre*, meaning to undertake) is one who supplies a good or service to the market for profit, often taking on the investment risks.[47] Such initiative is regarded as a significant ingredient of economic health and prosperity in advanced capitalist economies; indeed, many governments have focused on creating conditions in which such entrepreneurial 'spirit' can thrive and be rewarded. The virtues of entrepreneurship have their genealogy in modern liberal political thought and corresponding market society; rather than repressing individuals' destructive 'passions' in the face of some sovereign authority, it became increasingly efficacious to channel these energies into one countervailing, beneficial passion, which came to be known as the 'interests', and, crucially, economic interests, as Albert O. Hirschman details.[48]

In urban studies, the entrepreneurial city refers to an admixture of urban governance and private capital that thrives on inter-urban economic competition.[49] Though it is arguable that cities have always been centres of wealth creation and competitive political-economics, the relationship between urbanism and consumption harbours a long history of conflict – often outright opposition. In the entrepreneurial cities of the present, however, this relationship is compelled to be complementary and harmonious.[50] These societies of spectacle now include naked profit-oriented place-making investment by 'place entrepreneurs'.[51] But what are the key facets of such entrepreneurial subjectivity, and how do they operate at the level of everyday life in cities?

Incorporation

To incorporate means to unite already-existing entities into one body; in the legal sense, this means to form a corporation. The history of the corporation in Western legal theory is crucially intertwined with the transition from feudalism to capitalism and consequent shifts in the mode of production from agrarian to industrial. In medieval and early-modern Europe, the most common invocation of the legal corporation was exercised in the constitution of municipalities.[52] The urban corporation was both a form of society and also a 'free' republic.[53] As Philip Stern explains, many of these early urban corporations, including the City of London, claimed privileges through custom rather than formal charter. This form of incorporation, as a legal and political institution, laid claim to both property and popular sovereignty through associated rights and privileges.[54]

For Hegel, a critic of liberalism, corporations were occupational associations capable of mediating between civil society and the state and combating the atomization of market society.[55] As 'collective, rather than atomic, and professional, rather than residential, bodies' corporations could also provide a sense of purpose, where individual agents and their subjective freedom found common configuration.[56] However, early 19th-century jurisprudence on this matter confronted a dilemma: recognizing the rights of the city as an exercise of free association frustrated the liberal preference for allocating power between the state and the individual.[57] Cities lost their strength of association and became subordinate to states – the vocation of municipal self-determination was progressively and effectively restricted to the protection of city and private property.[58]

More recently, studies on corporate–institutional relationships as proponents for neoliberalization have focused on transnational financial firms and their crucial role in privatizing infrastructure and real estate – with specific implications for cities.[59] The entrenchment of corporate power has been correspondent with vast obfuscation and the rise of so-called 'immaterial labour'; the factory has become socialized and invisible.[60] Efforts at self-incorporation are especially encouraged for the precariat, which is compelled to celebrate the flexibility of 'self-employment'. The erosion of power of association in cities as corporate bodies made up of different individuals is staggering when cast against the power of private corporations in real estate and planning; 'public' infrastructure projects typically include contracts with numerous consulting firms, all billing hourly fees. Yet, efforts to mimic something like the popular sovereignty of early modern cities abound. The platform Nextdoor is an app that allows people to connect with their neighbours and share information and resources. It includes a handy 'Crime and Safety' category, which has, unsurprisingly, become a tool for neighbourhood policing.[61] Simple association alone will not combat atomization, but the history of conflict around cities as corporate bodies invites us to reexamine the rights of the city, especially as the neoliberally sanctioned quest for individual freedom leads – as Han argues – increasingly to its negation.

Surveillance

Surveillance as enabled through the built form and mechanisms of mass experience and mobility has transformed so that now, more and more, it is discreetly carried with us – concealed and commonplace. In exchange for seemingly boundless communication (increasingly a necessity) we exchange our personal information and preferences. Han writes: 'The neoliberal technology of power does not prohibit, protect, or repress; instead, it prospects, permits and projects'.[62] The prospecting of data is central to what Han calls 'the digital panopticon', an update of Bentham's system that 'thrives on its occupants' voluntary self-exposure'.[63]

The entrepreneurial individual and the entrepreneurial city meet on the plane of so-called 'smart' urbanism, where constant monitoring is slathered in a disarming discourse of positivity. Nothing looks like labour in the smart city, but surplus is constantly being generated. In the smart city, data = capital. In addition to the continued expansion of penal architecture – the 'over there' of the older panopticon – we have more intimate forms of control that also function, much like prisons, as capital-generating mechanisms. The figure of the entrepreneur and the notion of the entrepreneurial spirit still retain some relation to citizenship as an active, participatory, democratic practice; increasingly, however, the 'citizen' – yes, always a contested category – is replaced with the 'consumer', as firms like IBM, Siemens and Google intervene in urban governance and the provision of services, mining our data for value and then selling it back to us.[64] Smart cities now offer a new way for megalomaniac politicians and planners to test theories of social engineering, comparable to large-scale urban renewal projects of decades past. Hudson Yards, the largest private real-estate development in the history of the United States, as well as a testing site for smart urbanism, embodies the two urban imaginaries that characterized the tenure of Major Bloomberg: a corporate city, with the mayor as CEO; and the city as its own luxury product and playground for global elites.[65]

Of course, Bloomberg's 12-year tenure as Mayor of New York City also included the racist stop-and-frisk police program. Unfortunately, this program was merely a recent variation on centuries-long practices of racialized securitization in the city. In her work on the surveillance of blackness, Simone Browne considers how colonial New York City was 'a space of both terror and promise for black life'.[66] Browne examines the 'lantern laws' implemented in 18th-century New York City that required any unattended slave to carry a lit candle at night, part of the history of panoptic surveillance that demonstrates, as Browne argues, how race is integral to this history and to our present:

> Put another way, rather than seeing surveillance as something inaugurated by new technologies, such as automated facial recognition or unmanned

autonomous vehicles (or drones), to see it as ongoing is to insist that we factor in how racism and antiblackness undergird and sustain the intersecting surveillances of our present order.[67]

Browne traces surveillance from the slave ship to contemporary facial recognition software and biometrics. As a counterpractice, she takes up *sousveillance*, the ability of people to access and collect data on their surveillance and thereby neutralize or obfuscate it.[68] Browne proposes 'dark sousveillance' as 'an imaginative place from which to mobilize a critique of racializing surveillance, a critique that takes form in antisurveillance, countersurveillance, and other freedom practices'.[69]

Optimization

In computer science, optimization refers to the manipulation of a data set in order to reduce the number of steps in a formula while still producing the same outcome. More recently, optimization has been invoked to refer to self-actualization and the related industry of quantifiable progress toward some desired ideal. The age of Big Data ushers in a new version of the 'Quantified Self', with the motto 'Self Knowledge Through Numbers'.[70] A seemingly endless number of options abound to measure, monitor and enhance performance – physically, mentally, emotionally – and yet, we might argue that despite these endless manipulations of choice, the outcome is the same.

Establishing goals for the self becomes a way to shift attention away from the more obvious sites of labour and capital, and channel it into individualized management. In the mid-20th century, the principles of Taylorization developed for manufacturing industries became popularized in application to domestic management, as the home came to be regarded as a manufacturing site of a smaller scale, with new technologies and products introduced to increase efficiency.[71] In her study of modernization in France, Kristin Ross suggests that in this period from the mid-1950s to mid-1960s, efforts that once went into maintaining the colonies became concentrated on the level of everyday life in metropolitan existence.[72]

Writing in 1977, Barbara and John Ehrenreich conceived of the 'professional-managerial class' as having a major role in the reproduction of capitalist culture and class relations.[73] Reflecting on the fortunes of the professional-managerial class in the present conjuncture, Barbara Ehrenreich notes the importance of the service ethic to working people of various class positions. What happens when the service ethic is consistently underappreciated and, significantly, under-valued? What provides purpose? One indication is the obsessive turn to self-management techniques as a way to distract from the proliferation of what David Graeber calls 'bullshit jobs'.[74] 'Don't forget to invent your life', said Foucault in the early 1980s.[75] While we might interpret this as similar to the project of neoliberal entrepreneurship, we could also view it in a different register, more akin to what Graeber calls 'prefigurative politics': that is, 'making one's means as far as possible identical with one's ends, creating social relations and decision-making processes that at least approximate those that might exist in the kind of society we'd like to bring about'. For Graeber, revolutionary subjectivity consists above all in 'the defiant insistence on acting as if one is already free'.[76] For Han, hope lies rather in the figure of the '*idiot*' – not the one Marx and Engels alluded to in the *Manifesto*, but the one who can escape the 'compulsion to confirm' to the ubiquitous injunctions of 'freedom' and 'self-creation' curated by the spectacular technoculture of contemporary urbanism.[77]

Conclusion

The historical lesson seems straightforward: cities have for long offered freedom but also delivered tyranny. Like the development of capitalism, the process of urbanization is therefore to be seen as an inherently contradictory one, harbouring both emancipatory and alienating possibilities. Politically, this obliges us to consider the country, the city and the process of urbanization all together as a battleground, one inextricably tied to the manifold struggles over the reproduction or sublation of capitalism as we know it. In so doing, we draw special attention to the dialectical co-constitution of cities and citizens, examining especially the relationship between the production of space and the production of ideology. We have argued here and elsewhere that urban studies can usefully draw from several strands of critical theory to grasp the role of urbanism in producing political subjects and forms of subjectivity, on which the possibilities of cities beyond capitalism would seem to hinge. In this regard, we draw particular attention to the ideology of individualism informing long-standing liberal conceptions of freedom as much as manifestations alienation and atomization typically associated with capitalist cities, while underlining its latest mutations in the current moment of neoliberal urbanism.

A critical appreciation of Michel Foucault's work on subjectivity and Byung-Chul Han's recent theorization of 'psychopolitics' – undertaken with broad reference to theories of reification, commodification, culture industry and ideology – then allows us to sketch three strategies by which the subjective liberation marketed by urban techno-topias of big data and social media turns in fact into a negation rather than an actualization of freedom: incorporation, surveillance and optimization. While Foucault urges us to 'invent our lives' with 'technologies of the self', and Graeber advocates living our lives 'as if' we are already free, we end with Han's provocative invocation – in a manner reminiscent of Adorno's *Minima Moralia* and negative dialectics – of the figure of the 'idiot' as the antidote to the 'smart' and 'creative' subjects occupying the driving seats of neoliberal urbanism.[78] For this 'idiot is a modern day heretic', who 'stands opposed to the neoliberal power of domination', with the 'courage to deviate from the orthodoxy' of neoliberal urbanism's 'total communication and total surveillance'. As 'thoroughgoing digital networking and communication have massively amplified the compulsion to conform', it is the '*heretical consciousness*' of the 'idiot' that 'commands *free choice*'.[79]

Notes

1 Federici 2004:30–31.
2 Pirenne 1969/1925; Weber 1966/1927.
3 Pirenne 1969/1925; Weber 1966/1927.
4 Debord 1995/1967, §176, p.124.
5 Berman 1988/1982.
6 Engels 1993/1845.
7 Engels 1993/1845:36.
8 Ibid., pp.36–37.
9 Ibid., p.37.
10 Engels 1993/1845:37.
11 Marx and Engels 1998/1845:38.
12 Ibid., p.72.
13 Marx and Engels 1998/1845:72.
14 Marx and Engels 1998/1848:40.
15 Ibid., p.61.
16 Berman 1988/1982:13.
17 Ibid., p.15.
18 Ibid., p.24.
19 Ibid., p.24.
20 Williams 1973:1.
21 Jameson 1990:47.
22 Jameson 2018.
23 Brenner & Schmid 2014:60–163; Angelo 2017:164.
24 Williams 1973:1.
25 Hegel 1991/1821.
26 Lefebvre 1968, 2003/1970, 1991/1974.
27 Fanon 1963/1961; Abu-Lughod 1989; Abu-Lughod and Hay 1977; Hayden 1981; Massey 2005.
28 Marx 1963/1852.
29 Marx 1963/1852:123–124.
30 Hobsbawm 1998:11–12.
31 Lefebvre 2003/1970.
32 Debord 1995/1967:121–122.
33 Engels 1993/1845:54–55.
34 Davis 1990; Sorkin 1992; Rosler 1998, 1981.
35 Lynch 1960; Althusser 1971/1969:121–173.
36 Lukács 1971/1923; Adorno and Horkheimer 1944; Gramsci 1971; Foucault 1977/1975; 2007/2004, 2010/2004; Deleuze 1992/1990; Butler 1990.
37 Weber 1966/1927; Lukács 1971/1923; Adorno and Horkheimer 1997/1944; Marcuse 1964; Foucault 1977/1975, 2007/2004, 2010/2004.
38 Han 2017.
39 Foucault 1982:777–795.
40 Foucault 1982:784.
41 Foucault 1982:785.
42 Foucault 2010/2004.
43 Zamora 2019.
44 Ibid.
45 Han 2017:28.
46 Ibid., p.6.
47 *Oxford Dictionary of Business and Management* 2009.
48 Hirschman 1977.
49 Jessop and Sum 2000:2287–2313.

50 Heatherington and Cronin 2008:1–17.
51 Ibid., p.4.
52 Stern 2017:21–46.
53 Ibid., p.24.
54 Ibid.
55 Anderson 1992:275–379.
56 Ibid., p.289.
57 Frug 1980:1057–1154.
58 Ibid.
59 Jones 2017:172.
60 Rosler 2011.
61 Hempel 2017.
62 Han 2017:38.
63 Ibid., p.39.
64 Mattern 2014.
65 Ibid.
66 Browne 2015:82.
67 Ibid., pp.8–9.
68 Ibid., p.21.
69 Ibid.
70 Han 2017:60.
71 Ross 1995.
72 Ibid., pp.106–107.
73 Ehrenreich 2019.
74 Graeber 2018.
75 Zamora 2019.
76 Graeber 2014:5.
77 Han 2017:82.
78 Adorno 1974/1951, 1973/1966.
79 Han 2017:83 (original emphases).

References

Abu-Lughod, J., 1989, *Before European hegemony: The world system A.D. 1250–1350*, Oxford University Press, New York, NY.

Abu-Lughod, J. & Hay, R., Jr. (eds.), 1977, *Third world urbanization*, Routledge, New York, NY.

Adorno, T., 1973/1966, *Negative dialectics*, transl. E.B. Ashton, Continuum, New York, NY.

Adorno, T., 1974/1951, *Minima moralia*, transl. E.F.N. Jephcott, Verso, New York, NY.

Adorno, T. & Horkheimer, M., 1997/1944, *Dialectic of enlightenment*, Verso, London.

Althusser, L., 1971/1969, 'Ideology and ideological state apparatuses', in B. Brewster (transl.), *Lenin and philosophy*, pp. 121–173, Monthly Review Press, New York, NY.

Anderson, P., 1992, 'The ends of history', in *A zone of engagement*, pp. 275–379, Verso, London.

Angelo, H., 2017, 'From the city lens toward urbanisation as a way of seeing: Country/city binaries on an urbanizing planet', *Urban Studies* 54(1), 158–178.

Berman, M., 1988/1982, *All that is solid melts into air*, Penguin, New York, NY.

Brenner, N. & Schmid, C., 2014, 'Planetary urbanization', in N. Brenner (ed.), *Implosions/ explosions: Towards a study of planetary urbanization*, pp. 160–163, Jovis, Berlin.

Browne, S., 2015, *Dark matters: On the surveillance of blackness*, Duke University Press, Durham.

Butler, J., 1990, *Gender trouble: Feminism and the subversion of identity*, Routledge, New York, NY.

Davis, M., 1990, *City of quartz*, Verso, London.

Debord, G., 1995/1967, *Society of the spectacle*, transl. D. Nicholson-Smith, Zone Books, New York, NY.

Deleuze, G., 1992/1990, 'Postscript on the societies of control', *October* 59, 3–7.

Ehrenreich, B., 2019, 'On the origins of the professional-managerial class: An interview with Barbara Ehrenreich', Press A (interviewer) *Dissent Magazine*, www.dissentmaga-zine.org/online_articles/on-the-origins-of-the-professional-managerial-class-an-inter-view-with-barbara-ehrenreich.

Engels, F., 1993/1845, *The condition of the working class in England*, ed. D. McLellan and transl. F. Kelley-Wischnewetsky, Oxford University Press, Oxford.

Fanon, F., 1963/1961, *The wretched of the earth*, transl. C. Farrington, Grove Press, New York, NY.

Federici, S., 2004, *Caliban and the witch*, Autonomedia, Brooklyn.

Foucault, M., 1977/1975, *Discipline and punish: The birth of the prison*, transl. A. Sheridan, Pantheon Books, New York, NY.

Foucault, M., 1982, 'The subject and power', *Critical Inquiry* 8(4), 777–795.

Foucault, M., 2007/2004, *Security, territory, population: Lectures at the Collège de France, 1977–1978*, ed. M. Senellart and transl. G. Burchell, Picador, New York, NY.

Foucault, M., 2010/2004, *The birth of biopolitics: Lectures at the Collège de France, 1978–1979*, transl. G. Burchell, Picador, New York, NY.

Frug, G., 1980, 'The city as a legal concept', *Harvard Law Review* 93(6), 1057–1154.

Graeber, D., 2014, 'Anthropology and the rise of the professional-managerial class', *HAU: Journal of Ethnographic Theory* 4(3), 73–88.

Graeber, D., 2018, *Bullshit jobs: A theory*, Simon & Schuster, New York, NY.

Gramsci, A., 1971, *Selections from the prison notebooks of Antonio Gramsci*, ed. G.N. Smith & transl. Q. Hoare, International Publishers, New York, NY.

Han, B., 2017, *Psychopolitics*, transl. E. Butler, Verso, London.

Hayden, D., 1981, *The grand domestic revolution: A history of feminist designs for American homes, neighbourhoods, and cities*, MIT Press, Cambridge, MA.

Heatherington, K. & Cronin, A.M., 2008, 'Introduction', in K. Heatherington & A.M. Cronin (eds.), *Consuming the entrepreneurial city*, pp. 1–17, Routledge, New York, NY.

Hegel, G.W.F., 1991/1821, *Philosophy of right*, ed. and intro. A.W. Wood & transl. H.B. Nisbet, Cambridge University Press, Cambridge.

Hempel, J., 2017, 'For nextdoor, eliminating racism is no quick fix', *Wired*, 16 February, www.wired.com.

Hirschman, A.O., 1977, *The passions and the interests: Political arguments for capitalism before its triumph*, Princeton University Press, Princeton, NJ.

Hobsbawm, E.J., 1998, 'Introduction', in K. Marx & F. Engels (eds.), *The Communist manifesto: A modern edition*, Verso, London.

Jameson, F., 1990, *Postmodernism, or, the cultural logic of late capitalism*, Duke University Press, Durham.

Jameson, F., 2018, 24 November, 11:04, @JamesonFredric, www.twitter.com.

Jessop, B. & Sum, N., 2000, 'An entrepreneurial city in action: Hong Kong's emerging strategies in and for (inter)urban competition', *Urban Studies* 37(12), 2287–2313.

Jones, A., 2017, 'The corporation in geography', in G. Baars & A. Spicer (eds.), *The corporation: A critical, multi-disciplinary handbook*, Cambridge University Press, Cambridge.

Lefebvre, H., 1968, *Le droit à la ville*, Anthropos, Paris.

Lefebvre, H., 1991/1974, *The production of space*, transl. D. Nicholson-Smith, Blackwell, Oxford.

Lefebvre, H., 2003/1970, *The urban revolution*, transl. R. Bononno, University of Minnesota Press, Minneapolis.

Lukács, G., 1971/1923, *History and class consciousness: Studies in Marxist dialectics*, transl. R. Livingstone, Merlin Press, London.

Lynch, K., 1960, *The image of the city*, MIT Press, Cambridge.

Marcuse, H., 1964, *One dimensional man: Studies in the ideology of advanced industrial society*, Beacon Press, Boston, MA.

Marx, K., 1963/1852, *The eighteenth Brumaire of Louis Bonaparte*, International Publishers, New York, NY.

Marx, K., & Engels, F., 1998/1845, *The German ideology*, Prometheus Books, Amherst.

Marx, K., & Engels, F., 1998/1848, *The communist manifesto: A modern edition*, intro. E. Hobsbawm, Verso, London.

Massey, D., 2005, *For space*, Sage, London.

Mattern, S., 2014, 'Interfacing urban intelligence', *Places Journal*, https://placesjournal.org/article/interfacing-urban-intelligence/.

Oxford Dictionary of Business and Management, 2009, Oxford University Press, Oxford.

Pirenne, H., 1969/1925, *Medieval cities: Their origins and the revival of trade*, transl. F.D. Halsey, Princeton University Press, Princeton, NJ.

Rosler, M., 1981, 'In, around, and afterthoughts (on documentary photography)', *3 Works: Martha Rosler*, pp. 61–93, Press of the Nova Scotia College of Art and Design, Halifax.

Rosler, M., 1998, *If you lived here: The city in art, theory, and social activism*, The New Press, New York, NY.

Rosler, M., 2011, 'Culture class: Art, creativity, urbanism, part II', *E-flux* 23, www.e-flux.com/journal/23/67813/culture-class-art-creativity-urbanism-part-ii.

Ross, K., 1995, *Fast cars, clean bodies: Decolonization and the reordering of French culture*, MIT Press, Cambridge, MA.

Sorkin, M., 1992, *Variations on a theme park: The New American city and the end of public space*, Hill and Wang, New York, NY.

Stern, P.J. 2017, 'The corporation in history', in G. Baars & A. Spicer (eds.), *The corporation: A critical, multi-disciplinary handbook*, Cambridge University Press, Cambridge.

Weber, M., 1966/1927, *The city*, transl. D. Martindale & G. Neuwirth (eds.), Free Press, New York, NY.

Williams, R., 1973, *The country and the city*, Oxford University Press, Oxford.

Zamora, D., 2019, 'How Foucault got neoliberalism so wrong', interview by transl. K. Boucaud-Victoire, & S. Ackerman, *Jacobin Magazine*.

3 Can urbanization reduce inequality and limit climate change?

William W. Goldsmith

Introduction

To combat climate change and diminish social inequality, cities must do their part. They should be energy-efficient, comfortable and enjoyable, attractive, traversed by transit, their neighborhoods sprinkled with parks. Households should not have to double up or triple up in rental housing, and no one should have to sleep on the streets. Children should attend well-funded schools, to learn the skills to prosper and the ethics to participate in society. Everyone should pay a fair share of taxes.

The world's cities do not work this way, except in rare cases. Most cities abuse the climate and intensify inequality. Most do not offer quality schools for all children or affordable housing for all residents. Inequitable land-use rules aggravate climate change, regulatory failures diminish productivity, and austerity policies deepen social inequalities. Over the last half century, especially, cities and their regions have played a major part in intensifying inequality, damaging the environment, encouraging overconsumption, and requiring the overuse of fossil fuels. As one critic writes, our cities make it "nearly impossible to live lightly on earth."[1]

This chapter makes three arguments. As part of its program to limit climate change and reduce social inequality, the world needs to make drastic changes to dominant patterns of urbanization. Appropriate physical and social changes to city and metropolitan patterns will encounter powerful resistance. The politics of urban reform thus constitutes one of the key challenges of our time.

In the weeks since this essay was drafted in early 2020, COVID-19 has delivered catastrophic harm to the health and livelihoods of millions. Not only has the pandemic intensified inequalities and drawn attention to climate change. Perhaps of more significance, it has revealed the depth of these problems, which were with us before the pandemic and will remain when it subsides. (Comments or notes added after April 2020 are marked throughout by italics.)

Global urbanization

Current patterns of urbanization are not merely problematic. They are unsustainable. Badly constructed cities damage the environment, threatening global sustainability. They accelerate climate change, thus destroying species, pushing up sea levels, and intensifying storms, droughts, and floods. Global warming causes more frequent and intense wildfires, food shortages, disease epidemics, and human migrations. These events cause political stress as they worsen existing inequalities. In turn, the rising inequalities intensify damages to nature. Feeling entangled in such cruel feedback loops, anxious and poorly informed people come together in groups that act with suspicion or hostility. These groups cohere irrationally by ties of proximity, ethnicity, race, religion, or nationality. Threatened by poverty and environmental woe, frightened people compete in fruitless attempts to protect or to gain real or imagined advantages.[2]

The world resists acknowledging that the environmental imperatives are literally beyond human control: the earth's demands *will* be met, either by human design or by default. We will have either a sustainable world or environmental disaster. The basic science works with an inexorable logic that involves society, politics, economics, and technology: societies whose political systems allow deep inequalities will create biased economic systems. Such economies will invent and adopt environmentally abusive technologies. The world will not survive the abuse.[3] To turn around, to reverse the trajectory, to accept global environmental and economic limits, the world needs accommodating technologies, called into place by more sensible economic systems, which must be established by a changed politics. One major contributor will have to be more sustainable and equitable urbanization.

The world's processes of urbanization reinforce the tragic sorts of politics, economics, and technology that take us in the wrong directions ecologically and socially. It behooves us to ask how properly planned urbanization might help.

City growth and energy use

Urbanized populations have grown rapidly in the last half century. City regions now constitute majorities in most nations. Two-thirds of the world's population will live in cities by 2050.[4] At night these city regions beam bright images from satellites, indicating massive energy use. The images reveal ever higher levels of production. Seemingly limitless production calls forth ever-expanding consumption, in a dangerous cycle with associated waste, despoiling of air and water, depletion of resources, and warming of the atmosphere.[5]

Like most human activities, city-supporting industries and city living require energy, for transportation and construction, building use, and production of all sorts, including cooling for the internet, cloud storage and other high-tech needs, and agriculture. To supply that energy, we extract and burn fossil fuels, spewing into the atmosphere the CO_2, methane, particulate matter, and other pollutants that overwhelm what once was a resilient climate. To move away from dangerous reliance on fossil fuels and toward renewable sources of energy we must improve city patterns.

More appropriately designed cities can combat global warming. They can better utilize the advantages of density and social cooperation, to help restrain overproduction and overconsumption, and to diminish *per capita* levels of energy and resource use. Since urbanization has long been associated with falling fertility rates, it is highly likely that city patterns that result in reduced *per capita* production and consumption would be accompanied also by lower population growth rates, thus reducing *overall* energy use, reaping huge rewards.[6]

Cities, neoliberalism and social inequality

Badly designed cities facilitate inequality, distressing the losers, physically and emotionally, burdening political systems, corrupting the winners. Metropolitan regions have increasingly subdivided into distinct residential zones that separate rich from poor. A relatively small number of city regions have thrived, further separating rich from poor. Misshapen patterns of urbanization, with inequalities internal and external, feed a politics of austerity, provoking political tensions, promoting incivility, ultimately undermining democracy. In many ways, this situation is not sustainable.

The modern metropolis of New York is an example. Wealthy residents and many in the middle classes live in well-served neighborhoods in the giant city itself and in the suburbs of Westchester County, Long Island, New Jersey, and Connecticut. The poor live separately in poorly served neighborhoods in the city and suburbs alike. At one extreme, Greenwich, Connecticut, a small town just outside the huge city, is an icon for unsustainability. A haven for the uber-wealthy, it sits at the edge of one of the two or three richest suburban areas in the entire United States. Among this suburban area's one million inhabitants are 15 billionaires.[7] Greenwich itself, with only 13,000 residents, is home to corporate titans and real-estate moguls, and it houses nine of America's 25 richest hedge-fund managers. On average each earns $207 million a year.[8]

Greenwich has long been home to wealthy conservatives, but over the past half century, as the American economy has shifted from production to finance, these residents have moved far to the right, drawn to arch-conservative ideas rejecting the whole notion of government, resisting taxation, and questioning "the legitimacy of the liberal state." In contrast with earlier, more politically moderate residents, many in today's Greenwich believe, if only to justify or protect their own extraordinary wealth, that societies are and ought to be arranged not by public decisions but by individual actions. They like to believe that success and wealth result almost solely from private skill and hard work. They don't acknowledge the inheritance and luck that often play such an important part.[9] They condemn ideas of the Democratic Party – one prominent man remarks "that's how the U.S. will get to socialism – increasing government regulation." And they reject the old fashioned Greenwich ideas of Republicans who supported Dwight Eisenhower, President in the 1950s, or Nelson Rockefeller, New York State's governor in the 1960s and early 1970s. Leaders then thought government was "a working tool that should be used to shape society." They now support think tanks that advocate "low taxes and small government." The top Greenwich elected official, speaking at a Tea Party protest at Town Hall, said that "liberty has contracted today because the role of government has expanded."[10]

The idea of contrasting a place that is socially sustainable and fair with a place that is unsustainable and unequal is an enduring one. Plato, in the *Republic*, imagines two Greek societies. His proper city state is healthy. It sustainably limits consumption to meet actual needs. His luxurious city state knows no limits, living "beyond its needs in a perpetual quest for more."[11] It is unhealthy, and, millennia before human activity would threaten the globe's physical environment, it was unsustainable because the city's wealthy residents depended on unfair exploitation of the poor. One wonders what Plato would say about Greenwich.

Over the last half century, as the wealthy have self-isolated in increasingly exclusive suburbs and in growing numbers of gated "fortress" communities, geographic exploitation has become a reigning governing ethic.[12] Elites, protected ever more by spatial isolation, enjoy benefits without much notice of the costs, all the easier to ignore Plato's notion of the socially sustainable, healthy city or society. Elites comfortably imagine a society made up of autonomous individuals, all of whom somehow sustain *themselves*. Using their economic fortunes and political power to protect themselves, they ignore neighborhoods of the needy. The ultimate and absurd expression is the distant island estate, to which the uber-wealthy have been arranging in order to escape turmoil, from civil strife to COVID-19.[13]

Enthusiasm about "free" markets and efforts to erode government have serious consequences for city and metropolitan affairs. In the increasingly urbanized world, national and even international affairs often have a sizeable city imprint. When neoconservatives, libertarians, and others working on behalf of those with privileges aim to reduce regulations, cut taxes on business and the wealthy, and abandon redistributive spending, they damage city housing, transit, education, and health care.[14]

Exaggerated inequality has become more commonly accepted among elites, not only in city politics. National-level justifications of narrow, selfish interests are now common. Decades ago Margaret Thatcher put such ideas at the center of her political philosophy, infamously asserting that "there's no such thing as society."[15] People should not rely on collective action. Citizens acting as individuals or families should solve problems.

In line with such radically conservative ideas, city planning theorist Peter Hall proposed urban "enterprise zones." Celebrating the alleged advantage of an unregulated economy, Hall asserted that poor inner-city zones in Britain could be made to flourish like Hong Kong.[16] And when Ronald Reagan was elected president of the United States, in 1980, he expressed similar urban austerity thoughts, misleadingly claiming that enterprise zones would help cities, despite his hidden aim to weaken workers and unions while benefitting investors.[17]

Deregulation, austerity, and anti-city policies, the policies of the neoliberals, the unsustainable politics of the uber-wealthy arch conservatives, have caused grave damage. Their premise is wrong. Societies do exist. Individuals and households need to cooperate, ever more so as the highly urbanized world becomes more complex and interconnected. Individuals acting alone cannot deal effectively with climate change. Nor can individuals acting independently resolve inequality. An austere public policy disables government. A policy that seeks to minimize tax revenues disables the very agency, government, through which individuals can cooperate. Absent public coordination, individuals, families, and communities cannot function adequately in a world so intertwined. On the contrary, to reduce inequalities (and to limit climate change), societies need to impose regulations, coordinate markets, and restrain the political power of the wealthy. Protected suburban enclaves, such as Greenwich, and places such as the gated "fortress" communities that have proliferated on the edges of metropolitan areas across the United States, are designed to enhance such private power.[18]

Societies cannot thrive if they neglect their weakest members' essential needs, needs that include but go well beyond public safety and contract guarantees. Complex, specialized societies need to guarantee all members such basics as potable water, nutritious food, health care, shelter, education, child care, and means for accessibility. Kleptocratic regimes, of the sort which have recently proliferated across the globe, do not meet these global requirements. People need regimes that act to guarantee a tolerable existence for all.

Theory and experience tell us that organized cities can help. They can maintain health as they enhance accessibility, offer more housing, and support good schools. Twenty-five years ago sociologist Manuel Castells reported that West European cities did exactly such good things, as special places "of prosperity, peace, democracy, culture, science, welfare and civil rights," but he saw their qualities threatened on their "fragile island" in a "troubled world."[19] Politics and public policy created these beneficial urban possibilities, as residents worked to "impose their interests and their values."[20] More recently, in 1917, economist Angus Deaton explained why European mortality rates today are so much lower than American rates:

> The obvious difference is that the safety net is enormously more generous in Europe. And [a] lot of people in their 50s who lose their jobs can go on retirement. You get a doctor's certificate and you get paid pretty much your salary until you die.[21]

Today, public policy could organize cities – and societies – to reduce inequality and also to conserve energy, reduce consumption of resources, and minimize useless waste.[22]

The opportunity for survival through a simultaneous challenge to climate change and inequality is there, but it must be taken. Urbanization, done properly, should be one part of that endeavor. Since the world's societies are increasingly organized in cities or metropolises, then the advantages offered by those dense agglomerations need to be seized, their disadvantages minimized. Better cities can make life better for individuals, but only if good government and sensible economic policy respect technological and ecological imperatives.

In the COVID-19 crisis so far, public policy failures in the United States have revealed deep damages done by neo-liberal hostility to government. Well before the Trump administration, before his preposterous allegations about "fake news" and the "deep state," neoliberals had already worked to undermine the authority of public agencies, scientific findings, and social cooperation. Trump's hostility backed by the cowardly Republican Senate majority has proved literally fatal for many. The disproportionate suffering of black, Hispanic, other minority, and poor Americans has been revealed in their sharply higher illness and death rates.[23]

The tragic ineptitude of initial US responses to COVID-19 illustrates both the fragility of such "islands of peace and prosperity," as well as the dangers introduced when governments serve mainly the wealthy and their corporate interests, while ignoring the social and economic needs of ordinary citizens and flouting the ecological imperatives of physics and biology.[24]

To restate: The globe is beset by gargantuan threats of climate warming, inequality, and conflict. The globe is rapidly urbanizing in ways that worsen these threats. Achievable, alternative patterns of urbanization would help both ecologically and socially.

A half century of changing ecology and inequality

The last half century has been tumultuous environmentally. As challenges have intensified, recognition of the danger has grown deeper and more widespread. The challenges were long coming, as atmospheric greenhouse gases accumulated over centuries of resource use, coal at first, then petroleum and chemicals, plus deforestation, soil destruction, air and water pollution, and species decline. Resource and energy use have expanded since the beginnings of modern industrialization, massively and radically in recent decades. The trajectory is not sustainable.

Social, political, economic, and technological organizations have been slow to respond, despite warnings. Rachel Carson published *Silent Spring* nearly 60 years ago, in 1962, warning of life-endangering environmental abuses, especially chemical poisoning. Since then, scientists have warned industrialists, corporate leaders, and the public about poisons, waste, energy overuse, and greenhouse gas production. Warnings have grown more urgent with each subsequent decade, but rarely have governments responded at the necessary scale. Instead, parties with selfish interests, including corporate leaders, have pushed environmentally fraudulent theories together with an ideology promoting unlimited growth, individualism, limited government, low taxation, and public austerity. Some leaders even deny that human activities cause climate change, either in blind ignorance or with cynicism and negligence. Many leaders advocate ideologies of neglect as solutions for all problems, as they push ignorantly for short-term benefits.[25]

Looking at the long run of the entire 20th century, the story with social inequality begins more positively. For a period of more than half a century, high and graduated taxation and generous public spending along with rising productivity led to dramatically *reduced* levels of inequality in many places across Western Europe, the United States, Japan, the Soviet realm, and elsewhere.[26] Urbanization itself was sometimes managed progressively. Taxation and public programs reduced poverty and otherwise enhanced the lives of majorities, especially with schools, pensions, health care, and child care, but also transit, housing, and other services. Some regions of the world were more successful than others, as were some groups. There were catastrophic reversals including world wars and geno-cides. Nevertheless, for the majorities that survived, during much of the 20th cen-tury the industrialized nations reduced domestic inequalities and widened access to a higher quality of life.[27]

Trends shifted starting in the 1970s, however, and elections sharply demarcated the change around 1980. Rising inequality followed after loosened regulation of corporations and bountiful tax cuts for the wealthy, policies by Thatcher, Reagan, Germany's Helmut Kohl, and then a parade of other neoliberal leaders. As histo-rian Sean Wilentz says, "We live in a world where supply-side economics, which was always a fraud, became a religion."[28] High levels of inequality became hard to ignore just when major environmental sustainability challenges forced them-selves on the world.

Now, as monopoly and oligopoly dominate the economy, finance overpowers production, and corporations escape regulation,[29] metropolitan geography rein-forces inequality and it worsens climate change. Problematic metropolitan pat-terns will prove extremely resistant to correction.[30]

Constructing the metropolis of social inequality

Failed urbanization occurs prominently in the United States, but even there, it is set against an earlier period of relative success. From the end of World War II until the end of the 1970s, incomes for most US income strata grew steadily, reducing inequality. The bottom 90% of earners didn't do as well as the top earners, but still as a group they kept pace with the overall economy. Since the 1970s, however, the distribution of rewards has shifted radically upward. Relative incomes have declined not only for the groups at the very bottom, but on average for everyone except the top 10%. As economics journalist David Leonhardt writes, the trajec-tory of the economy "no longer tracks the well-being of most Americans. Instead, an outsized share of economic growth flows to the wealthy."

There has been steady shrinkage since 1980 of incomes compared to economic growth for everyone but the top 10%.[31] As Leonhardt writes, "[i]n the mid-20th century, GDP and incomes for most Americans tracked each other closely. If any-thing, mass incomes grew more quickly . . . but over the last 40 years, incomes have fallen badly behind economic growth."[32]

Anne Case and Angus Deaton have shown that this expanding inequality has led, quite remarkably, to what they call "deaths of despair." Mortality rates, which had declined in the United States for decades, suddenly rose, especially for young white Americans facing collapsing employment, and then more recently, for black Americans, as well. There is a sharp gap separating those with high school diplomas or less formal education from those with college degrees. The former are dying in unprecedented numbers by suicide, drug abuse, and alcoholism.[33]

Historians now see the earlier period, when inequality was diminishing, as out of synch with the expected routines of capitalist economic development.[34] It is now clear that the period of income convergence was anomalous, a result of the

New Deal in the United States and social democracy across Western Europe.[35] Rising inequality has once again become the rule.[36]

Of particular concern to us, income inequality has expanded similarly *inside* U.S. metropolitan areas. As the incomes of top-earners have risen, especially in cities specialized in finance and technology, the incomes of ordinary workers have stagnated. Earnings for *most* workers have stagnated nationwide, even in the booming metropolitan areas. The new geography of inequality concentrates high-income earners in highly productive metro areas, while lower-income earners lose out nearly everywhere.[37]

Over the last 40 years this geography of inequality *inside* US metropolitan areas has deepened, steadily increasing the barriers that insulate neighborhoods of the very well-off, like Greenwich, but also moderately well off suburbs, from neighborhoods of the working class and of the poor, especially neighborhoods where African Americans live.[38]

Over this period, investment in infrastructure for poorer neighborhoods has been grossly insufficient, leaving deteriorating roads, bridges, buildings, water supply systems, and schools. Housing stocks have also deteriorated, or been gentrified, leaving severe shortages of affordable housing, especially in booming metropolitan areas. The extreme housing shortage and explosion of homelessness in wealthy California's two giant metropolitan areas, Los Angeles and San Francisco, testify to the failures.

In all countries, the assignment of different socio-economic groups to separate residential areas reinforces the politics of inequality. In the United States these separations are especially severe, resulting from three features of the nation's history: the enduring legacy of chattel slavery followed by Jim Crow violence and exclusion, and then persistent racism; the widespread use of the automobile during principle periods of urban growth; and the ideology of individualism combined with a long tradition of "home rule." From these same roots have grown residential segregation and separate and unequal schools. The resulting land-use and school segregation patterns are enormously damaging and widely acknowledged.[39]

Autos, highways, and suburban municipalities

The extensively divided-up metropolitan territorial arrangement that began to emerge in the United States even before the 1930s became more pronounced after World War II and then dominant, as prosperity supported ever more sprawled development. Metropolitan land-use patterns were made more sturdy during the nation's most energetic economic expansions, until today they have become nearly indestructible, despite their heavy environmental and social burdens.

Automobile production was central to the building of this metropolitan arrangement, linked to petroleum, housebuilding, production of glass, plastic, rubber, steel, coal, cement, concrete, asphalt, wood and other building materials, trucking, and home appliances. Federal programs supported massive construction of millions of suburban single-family houses, the 41,000 mile interstate highway system, and much other highway building, providing access to suburban areas while demolishing city neighborhoods.

As each metropolis grew, boundary lines of the central municipalities stayed fixed.[40] Neither municipal boundaries nor responsibilities were extended to match growing populations. In suburban areas, local governments proliferated to an astounding extent.[41] Today, among large metropolitan areas, Pittsburgh has 463 local governments, the most *per capita.*[42] The San Francisco Bay Area includes three large cities, San Jose, population over a million, San Francisco, nearly 900,000, and Oakland, above 400,000. Each of these cities has a separately elected government with a full municipal apparatus. In addition the area boasts 97 smaller cities and nine counties, 165 school districts, dozens of utility districts, and still other authorities operating bridges, airports, and other facilities. Such multiplicity of governing authority is not only common, but long-standing. As early as the 1950s, the New York area boasted some 1,400 governments.[43]

As urban peripheries have developed in each US metropolis, state laws have enabled numerous new local governments to be established. Most places either incorporate as independent suburbs or join existing small towns. Suburban juris-dictions elect mayors and councils, and each usually runs a police department, court system, and jails, builds and operates its own municipal water supply and sewer drainage and treatment plants, paves and repairs streets and sidewalks, and operates parks and other public works. Most crucially, as we shall see, public schools belong to separate districts, often very small.[44]

Higher incomes and lower costs in the suburbs

One outcome of this city expansion at the periphery has been disparity in facili-ties. Older facilities serve central neighborhoods – worn out streets, risky bridges, decayed school buildings, broken down houses and apartments, outdated electric distribution systems, missing optic cables, leaky or even lead-emitting pipes for water supply, and defective sewerage and treatment plants. Newer buildings and facilities, less liable to need repair or replacement, serve relatively new suburban developments.

In the United States, better-off families tend to live farther from the center of the metropolis. More privileged areas are located toward the periphery, the clas-sic American suburbs.[45] In city regions in much of the rest of the world better-off households have tended to command central locations, relegating the poor to the periphery. In Paris elegant central districts are surrounded by less privileged zones stretching out to the far periphery. In most of the world the word *periphery* con-notes an undesirable residential location. In the United States the word *suburb* most often connotes middle-class success or more.[46]

US municipal boundaries insulate suburban finances and obligations from those of the central city. Cities need more and costlier repair and replacement of infrastructure, and city residents are least able to pay. Infrequently, often because of central-city (near) bankruptcy, state governments intervene to consoli-date metropolitan-area services, or to seize and manage city operations – police departments, public works such as water and sewage treatment, transit, parks and recreation, and schools. State government interventions may be administered to the disadvantage of cities, as the case when Michigan's governor appointed emergency managers for seven school districts and cities, including Flint and Detroit. The emergency manager for Detroit's city government, a bankruptcy lawyer from the Washington, DC, area, knew little about Detroit but was handed "a role with extraordinary power, usurping control from local elected officials."[47] Notably, in such interventions by state governments, the cities or districts affected are often majority black.

Racial segregation

"Minority" residential areas in the United States are sharply segregated from white districts, zones, neighborhoods, and suburbs.[48] As suburbs developed, homeownership was severely restricted to white buyers. Minority buyers were kept out by racially biased federal rules denying subsidies, by local zoning regulations prohibiting construction of apartments, subdivision into small lots, or other methods of lowering housing costs, by exclusion by real-estate brokers and mortgage bankers, and by overt hostility of white residents.[49] Even today, governments, banks, real-estate firms, and rental landlords practice widespread exclusionary practices, despite legal prohibitions.[50] Racial segregation is especially severe for blacks, somewhat less for Hispanics, and less still for other "non-white" and immigrant populations. Ethnic, linguistic, religious, and racial residential segregation is known in other nations, but rarely are the resulting patterns such a dominant territorial characteristic as in U.S. cities.[51]

No other influence so controls the quality and opportunity of U.S. communities as racial segregation. Family incomes influence residential location much less. The high correlation of race with household income means that darker-skinned residential areas almost have lower incomes. Many minority suburbs are burdened with inferior housing, and they receive lower levels of municipal services. Infamously, such disadvantaged districts may drink from poisoned public water supplies, or suffer brutally unequal levels of police protection.[52] Many white households are poor, of course, and white residential areas also are sorted by income, but white segregation by income is minimal compared to segregation by race. (*As I make last edits on this chapter, in early June 2020, widespread protests against racism and police violence are wracking the nation. The geography of protest is predictable, powerfully underlain by racial segregation.*)

Public schools

The adverse consequences of residential segregation by income and race (or ethnicity, language, religion, immigration status) are greatly exacerbated, even dominated, by another American tradition, "home rule," especially as it affects public schools.[53] The U.S. Department of Education does not operate the way education ministries do in many other nations. No national rules and regulations govern schools.[54] Rather, schools are governed separately by each of the 50 states.[55] Generally speaking, wealthy states can better fund their school systems, to provide better facilities and pay more highly qualified teachers. But the state-to-state gaps are the much smaller part of the story.[56]

Nearly every state delegates the organization of schools to the smallest municipal jurisdictions.[57] Nationwide the 50 states have created more than 13,000 school districts. Although laws at the state level regulate teacher qualification, facility conditions, and many curricular requirements, each district operates independently to construct buildings and other facilities, select and supervise teachers, and hire district superintendents, school principals, and all other personnel. Each district assigns its students to one of its schools. As noted earlier, the San Francisco Bay Area, with a population of 7.6 million, has 165 public school districts.

Each district pays for its schools in good part through local property taxes. Typically, local school taxes account for about half the property tax bills, which constitute substantial parts of most residents' *total* tax payments.[58] Nationwide, districts on average fund nearly 30% of school costs, with states funding most of the rest. Voters in local elections must approve budgets and tax rates.[59] From district to district, variations in household incomes and property values exert potent influence on the quality of schools.[60]

As legal theorist Gerald Frug has pointed out, this arrangement means that *public* schools function as though they were *private*.[61] In well-off suburbs, residents pay taxes to fund schools that only their children can attend. Districts with wealthy residents and expensive real estate fund their schools handsomely. They hire more experienced teachers, pay higher salaries, assign fewer students to each classroom, offer more specialized subjects, in music, the arts, science, and literature, build better facilities including laboratories, auditoriums, and sports fields, organize exotic international field trips, and more. Districts with low-income households do not have the funds for such things. Nearly all large central cities suffer this way.

The result, which is everyday common knowledge, is a huge gap separating two kinds of public schools. Schools in big city districts, which in many cases serve hundreds of thousands of predominately black, Hispanic, or immigrant students, offer drastically substandard educations.[62] These schools enroll much higher proportions of low-income students with special needs, including language training and health support, which increases costs.[63]

Schools in well-off suburban districts offer top-quality educations. Suburban students score at top levels on international tests, right up with winners from Finland and other Western European nations, Japan, Canada, Australia, South Korea, and Singapore. Students in big-city and minority school districts score much lower down.

The quality of local schools is frequently *the main concern* when families with children, or planning to have children, choose where to live.[64] It is difficult to overstate the powerfully negative influence on American life that comes down from this combination of racial segregation in housing, small local governing jurisdictions, and local funding of schools. The result is massive underfunding of the neediest schools and districts, and assignment of the least experienced teachers and the most run-down facilities to the poorest children, who often occupy the most overcrowded classrooms.

To overcome this home-rule/school-budget/segregation dilemma, reformers for many years have advocated voluntary or mandatory sharing of students between or among poor and well-off districts, statewide rather than local funding, expansion of the number of highly skilled and well-paid teachers, and broader testing with higher standards. In rare cases where such efforts are amply undertaken, they succeed to create excellent opportunities in good schools. But the vast majority of big-city and poverty-district schoolchildren find few such opportunities. Most often, the only families in troubled districts to escape the weak schools are those fortunate enough to find the means, either by lucky enrollment in a magnet school, by moving to the suburbs, or by sending children to extraordinarily expensive private schools.

The metropolis and climate change

Human-caused emissions of climate-changing greenhouse gases, GHGs, threaten the earth.[65] In the United States, anthropogenic methane, an extremely potent GHG in the short term, emanates mainly (c. 70%) from natural gas and waste dumps.[66] Globally, the principle source of long-run GHGs is CO_2 emissions from fossil fuels, increased tenfold since the mid-1940s, already accounting for three-quarters of GHGs.[67] Particulate matter from diesel engines, black carbon, is a powerful warming agent, diminishing now in the developed world but rising elsewhere.

NASA's website on Global Climate Change asks: *Is it too late to prevent climate change?* The concise answer by NASA mentions three needs: (1) international coordination, (2) production of cleaner energy, and (3) dealing with metropolitan problems:

> [T]he solution will require . . . a globally-coordinated response (such as international policies and agreements between countries, a push to cleaner forms of energy) and *local efforts on the city- and regional-level (for example, public transport upgrades, energy efficiency improvements, sustainable city planning, etc.).*[68]

NASA is correct to name land-use and transportation policies among the principal causes of US environmental damage. EPA scientists and regulators have long known that sprawled land-use patterns cause pollution and overuse of energy. In the 1990s, EPA supported Smart Growth programs against sprawl.[69] But despite these nods by key agencies responding to scientific findings, public policy has essentially ignored land-use sources of climate change.[70] Federal and state laws and regulations have reduced emissions by mandating controls on emissions from automobile tailpipes and industrial smoke stacks, but they have not dealt effectively with problems of metropolitan unsustainability, ignoring profligate land-use patterns, ever increasing vehicle-miles traveled (VMT), runaway household consumption, and the need to shift from autos to transit, bicycles, and walking. Advocates of such changes find the imposition of federal or state land-use controls to be a daunting task. Just as suburban residents vote to protect benefits they derive from racial segregation and school-district separation, they vote also to protect their auto-based privileges.

Outside the federal government, in the research community, in some state houses, and in a good number of municipalities, citizen scientists, city planners, and metropolitan development experts have long pointed to the damaging climate effects of the runaway consumption and burgeoning energy use associated with low-density urban development, but to little avail. Their warnings do not resonate widely.

In April 2020, eight science reporters reviewed the state of affairs approaching Earth Day 50. In their comments on the urgency of point-source and mobile-source climate-change regulations to cut emissions from power plants and automobile tailpipes, they are insistent. But on the land-use changes needed to reduce VMT and building emissions, the science writers are hesitant if not silent. Even in the obvious case of wrong-headed federal subsidies for flood insurance, a nonsensical idea environmentally and economically which enables, encourages, and subsidizes home building in tidal plains or flood-prone areas, these science writers give the obstacles preventing better policies more ink than they give to the potential benefits.[71] In the metropolitan cases of concern to us, land-use patterns that expand consumption and virtually require the daily use of automobiles, the scientists' hesitation may derive from their realization that the practical, political obstacles are tougher, given the adjustments that would be required of millions of individual homeowners and other real-estate parties.[72]

Sprawl versus climate

Patterns of metropolitan development in the United States have moved exactly contrary to the requirements of a stable, healthy climate. Policies almost regularly lead in the wrong direction, each step bringing us closer to climate disaster. Without intending climate damage, indeed with little thought of environmental consequences at all, the nation permits or promotes environmentally reckless metropolitan development. Each separate case of development spawns little opposition because it is small, offers short-run benefits to many, and seems "natural." This process was particularly evident during the long period of prosperity. The result: ever more sprawling suburban development, ever more consumption, more damage to the climate.[73]

The period of almost persistently booming industrial and economic growth that locked in sprawl has been called The American Century, even though it lasted only three or four decades, from before 1945 until after 1975.[74] Taken as a whole, US suburban expansion entailed a persuasive cultural construction. The suburb, with an image that was neither dirty and dangerously urban, nor impoverished and hardscrabble rural, was romanticized by a mythical past, a pastoral happiness, celebrated by Hollywood to represent the American way of life.[75] But the reality, hardly so benign, kept out families with modest or low incomes, as well as blacks, Hispanic, and other "minority" families. For many years, the suburban pattern made it doubly difficult for women (envisioned as housewives) to enter the labor force.[76] The suburban reality was class exclusion, racial discrimination, and gender subordination. Perhaps even more damaging in the long run, the suburban reality also involved unsupportable dependence on ever expanding production and consumption of goods and energy. The very same social, economic, and political forces that produced the suburbs not only solidified American inequality but also destabilized the global climate.

The public policy that built the suburbs centered on housing and transportation, and it promoted sprawl. Federal guarantees enabled banks to offer easy terms for mortgage borrowing. Promoted as early as the Hoover administration (1929–1933) with flyers promising "a car in every garage, a chicken in every pot," suburban homeownership after the New Deal became a social and political lodestar for decades.[77] The building of extensive single-land-use zones of single-family houses on large lots did create wealth for (white) working class and middle class Americans, but as it did so it left others far behind and moved all of us into an ecological dead end.

Overconsumption

The development of sprawl involved massive construction of roads, highways, and freeways surrounding central cities. This expansion promoted the dominance of automobiles, virtually eliminating accessibility by walking, biking, or transit, requiring as a practical matter at least one car for each household. There is now one passenger car for every two persons in the United States, man, woman, or child. Highway building was central to the entire American political economy, tied through the automobile industry's octopus-like links to a whole panoply of other industries, and the automobile is now central to the American psyche.[78] By the 1950s, with the exception of commuters by train or bus in parts of a very few metropolitan areas, hardly any successful working class, middle class, or better-off adult walked, bicycled, or used the bus, streetcar, or subway. Automobiles had become essential for work, shopping, and leisure.

Despite shrinking families, new home sizes expanded. Although family size declined by 9% and household size by 31% from 1940 to 2019, new home sizes more than doubled, from 1,177 to 2,657 square feet.[79] As builders constructed ever larger homes, households equipped them with increasing numbers of appliances. Consumption of energy and natural resources rose accordingly.

The growth of suburbs involves not only housing and transportation, but a whole range of individual, family and household-supporting goods. To manufacture these goods requires massive consumption of natural resources and high levels of fossil-fuel use. Both cause and consequence of this pattern of urbanization, American overconsumption taxes the world's resources. Taking far more than their proportional share, Americans use roughly five times their *per capita* share of global energy. US corporations produce excessively and US residents consume extravagantly. The style and pattern of US urbanization has formed an inseparable part of this excess and extravagance.[80]

No one should be surprised: the mantra of the American economy has long been growth, growth, growth. Growth is ultimately unstainable. How can the American political economy respond?

Sources of reform

Against this sobering state of affairs, in which we confront twin problems of intensifying inequality and deteriorating climate, can better patterns of urbanization offer part of the solution? Questions about metropolitan reform need to be asked not merely in the United States but throughout the world. As patterns of sprawled urbanization and elevated levels of production and consumption begin to afflict China, India, and other nations, atop already high levels in wealthy nations of the West, threats of climate change grow deeper. In fact, the earth *will not tolerate* large-scale imitation of US *per capita* abuse of energy and resources. The earth cannot sustain such super-wastefulness without grave consequences.

The upshot: if civilization is to survive, energy use and climate abuse must be contained. Above all, in the rich industrial countries fossil-fuel energy use must be reduced. Efforts to limit climate change will encounter stiff resistance. Overcoming that resistance will require, among other things, reductions of inequality. The metropolitan problems of the United States – including multiplicity of local governments, home rule for schools, and racism – magnify and maintain inequality. These problems also serve to divert political attention away from an even broader national source of both inequality and climate change – highly concentrated corporate power.

The increasing power of corporations in politics draws support from the built-in inequalities of the US metropolis. Metropolitan growth has endowed suburbanites not only with ownership of their schools, but a sense that they own the environment. "To move to the suburbs is to express a preference for the private over the public."[81] As they privatize their lives they weaken their connection to the broader public, resist intrusion by "outside" governments. Suburbanites can see "themselves as property owners and taxpayers who not only bought private homes and yards but also 'bought' limited government."[82]

Corporate power

In the contemporary world corporate power plays a strong hand in both matters of concern here, environmental sustainability/damage and economic fairness/inequality. Fossil-fuel corporations and their most direct beneficiaries play oversized roles resisting the politics needed to sustain the environment. Political struggles need finance, so money matters. *Forbes*, which proudly refers to itself as the magazine of the ruling class, reported in March 2019 that "each year, the world's five largest publicly owned oil and gas companies spend approximately

$200 million on lobbying designed "to delay, control or block" binding climate-motivated policy."[83]

Oil companies and related business groups cynically spend "millions of dollars on campaigns to fight climate regulations at the same time they tout their dedication to a low-carbon future."[84] In the half year from May to October 2019, oil companies and their lobbies spent $17 million on Facebook ads alone. In addition to their federal lobbying, US corporations spend state-by-state to influence policy. In Colorado, energy firms contributed $41 million in well-concealed campaigns to oppose regulations requiring land-use setbacks for drilling to extract natural gas by hydraulic fracturing, or fracking. By comparison, Colorado Rising, the grassroots organization in favor of these setback regulations, raised only $1.2 million. A spokesperson said "It felt like an attack on our democracy, where you have these corporations spending millions and millions of dollars . . . fighting citizens who are simply trying to protect their families."[85]

Similar financing and lobbying favors corporations over citizens across the spectrum, including in electoral campaigns in the United States, which cost millions upon millions of dollars. The un-equalizing effects of these biases not only favor the few who are better-off but also promote corporate interests as they sustain auto-based, consumption-oriented, fossil-fuel-dependent metropolitan sprawl.[86]

The same corporate interests that block sensible politics for protecting the environment, including wealthy benefactors of the petroleum industries and closely connected banks and investment funds, also block the reform politics that would mitigate inequality. Although the role of big money has never been something to ignore in US politics, its nearly totalizing effect has rarely been so great as now. Although it would make little sense to attribute either problem, climate change or rampant inequality, to the Trump administration alone, equally it is important to recognize how much worse and intractable politics have become because of the perfidy and incompetence of Trump, enabled and abetted by the majority leader of the U.S. Senate.[87]

Municipal reforms

What geographic level is most appropriate for fighting climate change, for limiting inequality? How much by international agencies, by the federal government, by states and municipalities, or by social movements? Should we hope that *municipal* reform or *state/provincial* reform of politics and policy making might work as a prime mover? Or should we focus on *national* changes, especially for fighting climate change but also for improving the distribution of income, wealth, and power?

Ada Colau, the progressive activist who was elected mayor of Barcelona in 2015, says that "city governments are key to the democratic revolution." But Colau quickly adds caution, that cities alone "cannot solve all the problems we are facing . . . because today the economy does not have borders."[88] *Nor, as the COVID-19 crisis reminds us, can municipalities or states print money.* To be sure, municipal economic boundaries are almost non-existent and municipal governments generally are very weak compared to nation states. Still, not only Spanish, or European, but U.S. experience also suggests the importance of city reform. But we need to know, as Colau's remark suggests: How strict are the limits to subnational public action?[89]

Since the Trump election in 2016, US federal agencies, such as the Environmental Protection Agency and the Department of Energy, have worked to withdraw and minimize interventions and regulations to limit global warming. Municipal and state authorities have responded to fill the gap, making promises and creating programs to reduce greenhouse gas emissions and fight climate change. Although it is far too early to judge the success of these local efforts, there is no doubt that they raise hope for lively political alternatives to the central government, perhaps as counterweights to corporate funding.[90]

America's Pledge constitutes a large group of U.S. cities along with states, businesses, and universities "committed to curbing global warming even as [the Trump] administration has walked away from the Paris Climate Agreement."[91] If *Pledge* members were a country, their economy would be the third largest in the world, bigger than all but two national signatories to the Paris Agreement. *The U.S. Climate Alliance* calls itself "a bi-partisan coalition of states . . . committed to the goal of reducing greenhouse gas emissions" to meet Paris goals. The 24 governors who are members of the *Alliance* represent more than half the U.S. population.[92] *We Are Still In*, another organization, which works together with *America's Pledge*, has membership comprising more than "3,500 representatives from . . . large and small businesses, mayors and governors, university presidents, faith leaders, tribal leaders, and cultural institutions." They come from all 50 states.[93] *America's Pledge* comprises 110 cities, 20 states, and over 1,400 businesses. They have adopted quantified reduction targets of nearly a gigaton of greenhouse gas emissions per year. They promote "renewable energy use and climate-friendly transportation systems" and promise to collect and disseminate data on subnational climate action that constitutes progress toward the U.S. pledge under the Paris accord. They aim also to encourage further climate action by cities, states, and business firms.

These groups know that local restraints on greenhouse gas emissions won't be sufficient in the face of opposition from Washington, which is lobbied massively by the fossil-fuel industry, but the groups nevertheless express optimism. Speaking in 2017, billionaire businessman Michael Bloomberg, the former New York City mayor, one of *America's Pledge*'s two conveners, said "there is nothing Washington can do to stop us." Jerry Brown, the other convener, when he was governor of the largest state, California, added that "states have real power as do cities," and that when they combine together with powerful corporations, progress is likely. Bloomberg says the group pledges "to continue to uphold our end of the [Paris Agreement] deal, with or without Washington."[94] Among other litigants, nine US cities, including New York and San Francisco, as well as the States of New York and Massachusetts, have sued fossil-fuel corporations over climate change, though success in the courts seems unlikely in the short term.[95]

As technological change advances, global temperatures continue to rise, and extreme and damaging weather becomes more common, more young people speak out and vote their interest in ecology and the environment. Demands for action are likely to expand, as alarms pop up on the screens of local progressives. In earlier efforts to force recognition of health and other damages from air and water pollution, the environmental justice movement brought together social activists with environmentalists. Today, evidence mounts that similar coalitions might combine environmental with social justice. But still, the obstacles are formidable. Winnie Byanyima, executive director of Oxfam International, said at the 2018 Global Climate Action Summit in San Francisco:

> Climate change is a political issue, not a technical issue. It's a question of justice and fairness. . . . It would take my grandfather, who lives in Uganda, 129 years to emit the same carbon as the average American citizen. . . . The emissions that are damaging our planet are being produced by the rich people, but the repercussions are hitting poor people the hardest."[96]

Despite the potent role that urbanization plays in creating or potentially stemming global warming, two questions remain: How much influence can local politics have? Even the largest of cities and states, even in coalition, face national and international corporate power, well-funded lobbying of politicians in Washington, and the overwhelming taxing power, regulatory power, and funding power of the federal government. And, even with the power, would local governments act to modify metropolitan land-use patterns, challenging the privilege and power of suburban establishments? This possibility seems remote.

Federal actions? Trump or a Democrat

As bad as the political situation for the environment has been under Donald Trump, it is sobering to recognize that his sharp deregulatory shifts, however unreasonable, are not unprecedented. Environmentalists and progressives have more to resist than the norm-breaking behavior of this president, even more to resist than the mean-spirited manipulations of the Republican Senate leadership. Political struggles over pollution, environmental regulation, global warming, and climate change, and also over social and economic inequality, have deeper roots. Damaging shifts in environmental policy from Obama to Trump have been drastic. But similar changes, even though less drastic, occurred also in 2001, when Republican George W. Bush replaced Democrat Bill Clinton in the White House.[97] A longer history shows that corporations and the wealthy persistently oppose inequality-reducing programs of taxation and spending, with reversals from one administration to another.[98]

Nevertheless, it is difficult to overstate the damage done by the Trump White House. In many key agencies, including the Environmental Protection Agency (EPA) and the Department of Energy (DOE), prominent investors, managers, and lobbyists have taken top positions, only to act in the short-run interests of their fossil-fuel businesses.[99] One can hardly imagine a clearer signal of attitude toward environmental affairs than the President's appointment of the head of fossil-fuel giant Exxon as Secretary of State. Other appointments have installed people equally hostile but much less capable.[100] Many federal department chiefs and their top deputies today are avowed enemies of environmental regulation. They include climate-change deniers and partisan, anti-science frauds.[101]

The Trump administration has openly assaulted science, even denying that climate change is occurring.[102] Most of the regulatory reductions are domestic, but they carry global consequences. The White House took the drastic step in 2017 of announcing its intention to withdraw from the Paris Climate Agreement, despite widespread support for the accord not only from the population at large but also from the business world.[103]

One cannot project, but only speculate, that should Democrats recapture the White House in 2020, and then confront inequality as well as the existential threat of climate change, the United States might move in better directions. The Green New Deal could become a reality. Still, while it is *possible* to imagine the implementation of federal policies to noticeably limit inequality, it is harder to imagine how sufficiently strong policies for housing, transit, and land-use could be put in place adequate to move the metropolis against climate change.

Notes

1 Marris 2020.
2 This chapter ignores a third existential threat, of international wars and nuclear attacks. Peter Marcuse (2019), 104–106, @ 105, asks, is "environmental degradation a greater danger in the long run than war, fascism, poverty, hunger, or disease?"
3 Barry Commoner (1976), the ecologist who ran for the U.S. presidency in 1980, laid out both the ecological and social cases for sustainable and more equitable urbanization in *The Poverty of Power*.
4 The United Nations Department of Economic and Social Affairs projects "68% of the world population . . . in urban areas by 2050." The figure was 55% in 2019. See United Nations 2018.
5 See Olpadwala and Goldsmith (1992). On choking versus breathing, see Gardiner 2019.
6 Urbanization reduces fertility rates, family size, and population growth. See Martine, Alves and Cavenaghi (2013) and, on consumption, Torrey (2004).
7 The Bridgeport–Stamford–Norwalk–Danbury metropolitan statistical area, which comprises all of Fairfield County, with one million residents, has the 2nd or 3rd highest average personal income in the nation. Greenwich has a population of about 13,000.
8 Reported by Osnos 2020. *Forbes* counts 2,000+billionaires in the United States.
9 On the heavy role of luck, and the winners' inability to believe that, see Frank 2016.
10 Osnos (2020). The remark on the working tool and small government was made by a man whose fortune came from digital trading. Notably, as Chamoff (2010) reports, among its more than 60,000 residents, the Greenwich *school district* has a sizeable poverty population. Of the 70% of the district's students who attended public schools in 2010, 38% were minority, mostly Hispanic.
11 DeWeese-Boyd & DeWeese-Boyd 2007 examine contemporary rich *nations* thus exploiting poor nations.
12 Blakely & Snyder 1999. Rodden (2019) shows the depth of geographic political divisions in the United States.

13 Peter Thiel, Silicon Valley's notable Trump supporter, fears "climate catastrophe, decline of transatlantic political orders, resurgent nuclear terror." O'Connell 2018; Post-Covid-19, see Flemming 2020.

14 "In Britain, in some key areas the state was not rolled back at the beginning. In Thatcher's first two terms there was more continuity than change in health, social services and education. People suffered spending squeezes, but nothing on the scale of the subsequently savage cuts to benefits and housing. It was not until her third term that restructuring began" (Thatcher 2013b). See Dean 2013.

15 British prime minister, 1979: "They are casting their problems at society. And, you know, *there's no such thing as society*. There are individual men and women and there are families. And no government can do anything except through people, and people must look after themselves first" (Thatcher 2013a).

16 See, e.g., Goldsmith 1982.

17 Goldsmith 1982. Helmut Kohl, elected chancellor of Germany in 1982, also pursued austerity (Marshall 1996).

18 Blakely & Snyder 1999.

19 Castells 1994.

20 Castells 1994: "Cities are socially determined in their forms and in their processes . . . played out, and twisted, by social actors that impose their interests and their values, to project the city of their dreams and to fight the space of their nightmares." On the complex roots of European social democracy, see Dorrien 2019.

21 Guo 2017.

22 Raworth (2017) proposes a "doughnut economy" to guarantee basic needs *and to limit* consumption by the rich.

23 On historical biases, see Perry, Harshbarger & Romer 2020. On current biases, see Flitter& Eavis 2020.

24 Jane Mayer (2020) explains how majority leader McConnell, despite his recognition of Trump's incompetence and self-dealing, serves (and makes) big money by enabling the President to stay in power, thus intensifying inequality and aggravating climate change.

25 The environmental record is poor in both capitalist and Communist states, as both pursue economic growth. See Olpadwala & Goldsmith 1992.

26 Putting aside the unequally distributed costs of devastating wars and the miseries of late colonialism, imperialism, and racial and ethnic oppression.

27 Reductions in inequality went further in Western European social democracies, Canada, and Japan than in the US limited welfare state. Despite widespread agreement that inequality is bad, there is little agreement in the United States even about how to measure it, less about how it should be reduced. See Rothman 2020.

28 Kolhatkar 2020.

29 I leave open how to characterize this period – late, monopoly, authoritarian, or kleptocratic capitalism, or something else. See Suarez-Villa 2015.

30 This chapter deals with American cities and society, but authoritarian threats to democracy are worldwide. For an abbreviated list consider Brazil, Turkey, Poland, the Philippines, and China. On demagoguery see Levitsky & Ziblatt 2018. Admittedly, to look at one nation, especially the United States, is to adopt a particularly narrow viewpoint. But the United States boasts the largest and most highly advanced industrial economy, with immense global influence. It hosts the world's most powerful military. Moreover, despite great historic flaws of imperialism, racism, inequality, and demagoguery since the 2016 elections, the United States may retain value as a symbol of the possibilities afforded by mobility, democracy, and diversity.

31 Leonhardt 2019 uses research and data from the Federal Reserve and from Piketty *et al*. 2018.

32 Leonhardt 2019.

33 Case and Deaton 2015, 2020.

34 Piketty 2014.

35 These gross trends are oversimplified, of course, and it is impossible to represent important changes with averages. One cannot ignore the effects of World War II. During the entire period, urban environments worsened nearly everywhere, and in cities in the emerging economies "urban environmental damage [had] serious, adverse consequences for . . . the quality of life, particularly for poor people." See Olpadwala & Goldsmith 1992.

36 But the rise of China, for example, reduces the economic disparities separating the former First and Third Worlds.

37 VanHeuvelen & Copas 2019; Badger & Quealy 2019a, 2019b. Whereas the ratio of wages for the top 10% of earners compared to the bottom 10% in 1980 clustered in most metros, ranging from 3.5 to 5.0, by 2015 those ratios had risen to between 4.5 and 6.0, but by much more in the largest cities. For the New York and Los Angeles metros, the ratios were below 4.5 in 1980, but at about 7.0 in 2015.

38 Chetty *et al.* 2019.

39 See, e.g., Warner 1962; Jackson 1985; Goldsmith and Blakely 1992, 2010; and Massey and Denton 1993.

40 Exceptions occurred, mostly in the South, where some central cities annexed peripheral communities. A notable North American case of authority once spreading by design to match the need was Metro Toronto.

41 In unusual cases, cities did incorporate suburban territories, mainly in the deep South, in an effort to prevent majority black populations in central cities from potential electoral victories.

42 Maciag 2019.

43 Wood and Almendinger 1961.

44 Even in the current period, cities can attend to social needs rather than only corporate profits. In the Detroit area, for example, as jobs disappeared, the government only extended unemployment benefits, while across the border, in Windsor, Ontario, "Canada emphasized job retraining, rapidly steering workers into new jobs . . . and Canadian workers also did not have to worry about losing health insurance" (Kristof & WuDunn 2019).

45 The detailed pattern in any particular metropolis is more complex, a palimpsest, like a checkerboard tablecloth overlaying a circular design, each annulus poorer with distance from the center. See "doughnuts and checkerboards," Goldsmith 2016:110 ff. Growing numbers of poor, minority suburbs are located outside city centers (Flint, MI, is an example, see the following), and in several booming cities high income residential populations are displacing poor residents (e.g. New York, San Francisco, Seattle, Washington), but even in those areas, higher-income suburban districts are still the rule. On the key role of public schools, see the following section.

46 Goldsmith 1997. Geographically intricate metropolitan areas now include impoverished and under-served suburban areas.

47 Bosman & Davey 2016.

48 See Massey & Denton 1993; Goldsmith & Blakely 1992. Oddly, as Hanchett (1998) shows, the main exceptions were in some cities of the South, before integration laws of the 1950s and 1960s, when *social* distinctions and segregation by race were so strong that whites did not demand *physical* segregation.

49 Green Lines and Red Lines outlined ineligible areas on real estate maps. For a vivid review of federal complicity, see Burns 2020.

50 According to the Urban Institute, "Five decades after the passage of the Fair Housing Act, and even after amendments . . . the United States still faces significant challenges to creating inclusive communities" (Hendey & Cohen 2017; O'Donnell 2020). Also see Goetz 2018.

51 For an international example, see Bou Akar 2018. Although US levels of residential segregation have diminished, they are still very high. Goldsmith 1997.

52 Regarding police brutality see Goldsmith 2015; Morse, Simpson & Stein 2020 and other works in Angotti 2019; Associated Press 2014. Ferguson, a black St Louis suburb; Evan Hill *et al.* 2020. Minneapolis. Regarding lead contamination of the water supply in the black Detroit suburban city of Flint, Michigan, see Holden, Fonger & Glenza 2019.

53 More than 90% of children attend public schools. Two percent attend private schools and 7–8% attend religious schools, mostly Catholic parochial schools, according to the National Center for Education Statistics.

54 There *are* federal requirements, which schools *do* meet in order to receive (relatively modest but still important) special funds.

55 And also by Washington, DC, and "territories," such as Puerto Rico, the US Virgin Islands, and Guam.

56 Federal funds do go to public school districts, with conditions, but in relatively minor amounts.

57 The degree of localism varies considerably state to state. New York State, for example, consolidated many very small school districts decades ago, but 950 still exist. Some – in New York City, Buffalo, and other big cities – are huge, but most are very small, extremely inefficient, and of highly varying quality.

58 Taxes are collected mainly as federal income taxes, sometimes state income taxes, state and local sales taxes, and local real property taxes. Few cities collect income taxes, usually at the 1% to 2% level; 41 states tax income.

59 Nath 2015. Arkansas provides more than 80% of school funding from the state budget, far more than any other state. About half the states provide half or less, including about a dozen that require localities to provide 60% or more through local taxes. Some states restrict local tax options. The most famous such restriction is California's Proposition 13, which through its limits on local tax levels has driven the state's schools from near the top of national rankings in the 1970s to near the bottom today.

60 Within big city school districts, parents in well-off neighborhoods sometimes privately subsidize their own neighborhood schools.

61 Frug 1999.

62 In one of the Bay Area's largest school districts, Oakland, with minority and poverty populations, a large-sample study of students in 2010 found that barely half, 53%, graduated on schedule, and an astonishing 37% dropped out of school all together. Goldsmith 2016:8.

63 The resulting average national scores drive US rankings way down. See Goldsmith 2016, chapters 3 and 4.

64 Ingram and Kenyon 2014.

65 Climate change is caused by black carbon (particulate matter, or soot, mainly from diesel engines) and greenhouse gases, or GHGs, mainly carbon dioxide and methane. NASA n.d.b Other GHGs are water vapor, nitrous oxide, 6% of global emissions, and fluorinated gases, now 2%, after limitations by international agreements.

66 Methane accounts for 16% of global GHGs with short-term warming effects many times more powerful than CO_2. Methane is emitted (worldwide) mainly from animal agriculture and oil and natural gas operations. For US sources, see U.S. Energy Information Administration 2011.

67 See Environmental Protection Agency n.d.

68 NASA n.d.a. Emphasis added.

69 Discussions took place on EPA's Clean Air Act Advisory Committee, led by Assistant Administrator Mary Nichols.

70 One exceptional case was the January 17, 1998, hold-up of hundreds of millions of dollars of highway subsidies to the Atlanta region. EPA applied the "Conformity" provisions of the Clean Air Act, allowing the cutting of DOT funding if a metro area exceeded air pollution limits. See Conformity 1998.

71 Notably, when the Trump White House did the right thing, calling for limiting such flood disaster insurance, the Congress beat back the initiative.
72 Science section, *New York Times*, April 21, 2020.
73 Sprawl is difficult to define precisely, but we know it when we see it. See Ewing, Pendal & Chen 2003.
74 Goldsmith & Blakely 2010.
75 On this pastoral myth, see Williams 1973; Marx 1964; Hayden 2003.
76 See Dolores Hayden 2003.
77 At least the first phrase was used by Hoover's campaign as a newspaper ad: Chicken 1928.
78 Del Mastro 2019.
79 US Census of Housing, Table HH-6. From 1973 to 2010, the average size of a new house in a metropolitan area increased almost 40%. The period for house size doubling was 1940–2014.
80 Elsewhere (Goldsmith 2016), I have argued that a more appropriate pattern of US urbanization could enhance schooling, lessen inequality, and offer other benefits.
81 Schneider 1992, cited in Starr 2019.
82 Starr 2019.
83 McCarthy 2019. According to Influence Map 2019, "BP has the highest annual expenditure on climate lobbying at $53 million, followed by Shell with $49 million and ExxonMobil with $41 million. Chevron and Total each spend around $29 million every year." Also see Jacobs 2016.
84 Laville & Pegg 2019.
85 The corporations are out-of-state and out-of-country. Laville & Pegg 2019.
86 Federal legislation regarding COVID-19 exhibits similarly pro-corporate and racist bias. Eisinger 2020; Eligon and Burch 2020.
87 Mayer 2020. While it preceded Trump's entry into politics, the Supreme Court's "Citizens United" decision in 2010 received key support from McConnell. (*It is possible that the multiple tragedies of COVID-19 may trip the balance, so that both Trump and McConnell will lose power. In that case, and especially because so much energy will go to rebuilding the economy, questions and answers are crucial if we are to reform to improve the distribution of income and wealth and to reverse global warming.*)
88 Colau 2015
89 For examples of local progressive action in the United States, see Clavel 1986, 2010.
90 *In the case of the Corona virus, strikingly similar contrasts separate the flailing, ineffective, and panic-inducing actions by the federal government from highly competent, reassuring responses by several state governments, including New York and California.*
91 Jordans & Thiesing 2017; America's Pledge n.d.
92 The largest seven of those states – CA, NY, PA, IL, NJ, VI, and WA – make up more than a third of the 2019 US population, but elect only one-seventh of the US Senate.
93 We Are Still In n.d.
94 Regan 2017. New evidence appears nearly every day of the push back. Governors in Florida and other Republican states, together with Democratic governors, resisted the Trump administration's 2019 proposal to resume long-banned off-shore oil drilling. Also see Miller 2017.
95 See Hasemyer 2020; Geiling 2018; Schwartz 2019.
96 Byanyima, quoted by Mendez 2020:204.
97 The United States joined the 1997 Kyoto Protocol to limit greenhouse gas emissions when Clinton was president only to withdraw when Bush became president. Industrial representatives on EPA's Clean Air Act Advisory Committee drastically reduced their interest in environmental regulation when the presidential shift occurred (personal observation).
98 See, e.g., Isaac Martin 2008, 2013, on resistance to the federal income tax and on California's municipal tax-limiting Proposition 13.

99   The replacement of politicians or professionals by lobbyists has been a hallmark of the Trump administration. Friedman 2018; Lipton & Ivory 2017; Friedman & O'Neill 2020.
100   The former CEO, Rex Tillerson, proceeded to dismantle and demoralize the State Department. He resigned or was dismissed in March 2018.
101   Lewis 2018.
102   For an extensive list of mostly deregulatory changes, see Greshko 2019.
103   The US withdrawal is scheduled to be official in November 2020. On business support see Center for Climate and Energy Solutions, n.d. Also see Cohen 2018. After inequality and climate change, there is a third existential threat, nuclear war. Trump would seem to have brought that closer as well, by withdrawing from the Iran multilateral treaty.

References

America's Pledge, n.d., www.americaspledgeonclimate.com/fulfilling-americas-pledge/.
Angotti, T. (ed.), 2019, *Transformative planning: Radical alternatives to neoliberal urbanism*, Black Rose Books, Chicago, IL.
Associated Press, 2014, 'Police fatal shooting of Black teenager draws angry crowd in St. Louis Suburb', *Associated Press*, 10 August.
Badger, E. & Quealy, K., 2019a,'The upshot: Big cities have prospered, but inequality has soared', *New York Times*, 2 December.
Badger, E. & Quealy, K., 2019b. 'Watch 4 decades of inequality drive American cities apart', *New York Times*, 2 December, citing Analysis of census and American Community Survey data by Jaison Abel and Richard Deitz, Federal Reserve Bank of New York.
Blakely, E.J. & Snyder, M.G., 1999, *Fortress America: Gated communities in the United States*, Brookings, Washington, DC.
Bosman, J. & Davey, M., 2016, 'Anger in Michigan over appointing emergency managers', *New York Times*, 22 January.
Bou Akar, H., 2018, *For the war yet to come: Planning Beirut's frontiers*, Stanford University Press, Stanford, CA.
Burns, S. & McMahon, D., 2020. 'East lake meadows: A public housing story', TV Movie, PBS Documentary, 13 January.
Case, A. & Deaton, A., 2015, 'Rising morbidity and mortality in midlife among White non-Hispanic Americans in the 21st century', *Proceedings of the National Academy of Sciences*, 8 December.
Case, A. & Deaton, A., 2020, *Deaths of despair and the future of capitalism*, Princeton University Press, Princeton, NJ.
Castells, M., 1994,'European cities, the informational society, and the global economy', *New Left Review* 204, March–April.
Center for Climate and Energy Solutions, n.d., www.c2es.org/content/business-support-for-the-paris-agreement.
Chamoff, L., 2010, 'With nearly 3 in 10 in private schools, public district looks to retain best and brightest', *Greenwich Time*, 30 November.
Chetty, R., Hendren, N., Jones, M.R. & Porter, S.R., 2019, 'Race and economic opportunity in an intergenerational perspective', *Quarterly Journal of Economics* 711–783, May.
Chicken, 1928, https://iowaculture.gov/history/education/educator-resources/primary-source-sets/great-depression-and-herbert-hoover/chicken.
Clavel, P., 1986, *Progressive city: Planning and participation, 1969–1984*, Rutgers University Press, New Brunswick, NJ.

Clavel, P., 2010, *Activists in city hall: The progressive response to the Reagan era in Boston and Chicago*, Cornell University Press, Ithaca, NY.

Cohen, S., 2018, 'The states resist Trump's environmental agenda', *Earth Institute*, https://blogs.ei.columbia.edu/2018/05/07/states-resist-trumps-environmental-agenda/.

Colau, A., 2015, 'Democracy now: The war and peace report', *NPR*, 5 June.

Commoner, B., 1976, *The poverty of power – Energy and the economic crisis*, Knopf, New York, NY.

Conformity, 1998, https://digital.library.unt.edu/ark:/67531/metacrs936/m1/1/high_res_d/RL30131_1999Oct15.html#Table%202.

Dean, M., 2013, 'Margaret Thatcher's policies hit the poor hardest – and it's happening again', *The Guardian*, 9 April.

Del Mastro, A., 2019, 'Car culture and suburbia in the American psyche', *The American Conservative*, 22 February.

DeWeese-Boyd, I. & DeWeese-Boyd, M., 2007, 'The health city versus the luxurious city in Plato's *Republic*: Lessons about consumption and sustainability for a globalizing economy', *Contemporary Justice Review* 10(1), March.

Dorrien, G., 2019, *Social democracy in the making: Political and religious roots of European socialism*, Yale University Press, New Haven.

Eisinger, J., 2020, 'How the coronavirus bailout repeats 2008's mistakes: Huge corporate payoffs with little accountability', *Propublica*, 7 April, viewed 15 April, from www.propublica.org/article/how-the-coronavirus-bailout-repeats-2008s-mistakes-huge-corporate-payoffs-with-little-accountability.

Eligon, J. & Burch, A.D.S., 2020, 'Questions of bias in Covid-19 treatment add to mourning for Black families', *New York Times*, 10 May.

Environmental Protection Agency, n.d., viewed 10 March 2020, from www.epa.gov/ghgemissions/global-greenhouse-gas-emissions-data#Trends.

Ewing, R., Pendal, R. & Chen, D., 2003, 'Measuring sprawl and its transportation impacts', *TRB 1831*, 1 January, p. 1.

Flemming, J., 2020, 'Bunker with a bowling alley: How the rich are running from coronavirus', *Los Angeles Times*, 23 March.

Flitter, E. & Eavis, P., 2020, 'The buybacks that ate restaurants' cash up', *New York Times*, Business Section, p. 1 ff, 25 April.

Frank, R.H., 2016, *Success and luck: Good fortune and the myth of meritocracy*, Princeton University Press, Princeton, NJ.

Friedman, L., 2018, 'New energy secretary fits trend: Cabinet dominated by lobbyists', *New York Times*, 18 October.

Friedman, L. & O'Neill, C., 2020, 'Who controls trump's environmental policy?', *New York Times*, 24 January.

Frug, G.E., 1999, *City making: Building communities without building walls*, Princeton University Press, Princeton, NJ.

Gardiner, B., 2019, *Choked: Life and breath in the age of air pollution*, University of Chicago Press, Chicago.

Geiling, N., 2018, 'The list of cities suing major fossil fuel companies over climate change just got longer', *Think Progress*, 23 January, https://thinkprogress.org/richmond-california-climate-lawsuit-478382e0d16e/.

Goetz, E.G., 2018, *The one-way street of integration: Fair housing and pursuit of racial justice in American cities*, Cornell University Press, Ithaca.

Goldsmith, W.W., 1982, 'Bringing the third world home: Enterprise zones', *Working Papers Magazine*, March.

Goldsmith, W.W., 1997, 'The metropolis and globalization: The dialectics of racial discrimination, deregulation, and urban form', *American Behavioral Scientist* 41(3), November–December.

Goldsmith, W.W., 2015, 'The drug war, prisons, and police killings of black men', *Progressive Planning* 203(spring), 25–28.

Goldsmith, W.W., 2016, *Saving our cities: A progressive plan to transform urban America*, Cornell University Press, Ithaca.

Goldsmith, W.W. & Blakely, E.J., 1992/2010, *Separate societies: Poverty and inequality in U.S. cities*, Temple University Press, Philadelphia.

Greshko, M. *et al.*, 2019, 'A running list of how President Trump is changing environmental policy', *National Geographic*, 3 May.

Guo, J., 2017, '"How dare you work on Whites": Professors under fire for research on White mortality', *Washington Post*, 6 April.

Hanchett, T.W., 1998, *Sorting out the new South city: Race, class, and urban development in Charlotte, 1875–1975*, UNC Press, Chapel Hill, NC.

Hasemyer, D., 2020, 'Fossil fuels on trial: Where the major climate change lawsuits stand today', *Inside Climate News*, 17 January. https://insideclimatenews.org/news/04042018/climate-change-fossil-fuel-company-lawsuits-timeline-exxon-children-california-cities-attorney-general.

Hayden, D., 2003, *Building suburbia: Green fields and urban growth, 1820–2000*, Pantheon, New York.

Hendey, L. & Cohen, M., 2017, *Using data to assess fair housing and improve access to opportunity a guidebook for community organizations*, Urban Institute, Washington, DC, August.

Hill, E., Ainara, T., Christiaan, T., Drew, J., Haley, W. & Robin, S., 2020, 'How George Floyd was killed in police custody', *New York Times*, 31 May.

Holden, E., Fonger, R. & Glenza, J., 2019, 'Revealed: Water company and city officials knew about Flint poison risk', *The Guardian*, 10 December.

Influence Map, 2019, *How big oil continues to oppose the Paris agreement*, https://influencemap.org/report/How-Big-Oil-Continues-to-Oppose-the-Paris-Agreement-38212275958aa21196dae3b76220bddc.

Ingram, G. & Kenyon, D.A. (eds.), 2014, *Education, land, and location*, www.lincolninst.edu/sites/default/files/pubfiles/school-quality-school-choice-residential-mobility_0.pdf.

Jackson, K., 1985, *Crabgrass frontier: The suburbanization of the United States*, Oxford University Press, Oxford.

Jacobs, M., 2016, 'America's never-ending oil consumption: Why presidents have found it so difficult to ask people to just use less**', *The Atlantic*, 15 May, www.theatlantic.com/politics/archive/2016/05/american-oil-consumption/482532/.

Jordans, F. & Thiesing, D., 2017, 'Defying Trump, U.S. cities and states pledge to support goals of Paris climate deal', *Los Angeles Times*, 11 November, www.latimes.com/nation/ct-paris-climate-deal-pledge-20171111-story.html.

Kolhatkar, S., 2020, 'Embarrassment of riches', *The New Yorker*, 6 January.

Kristof, N.D. & WuDunn, S., 2019, *Tightrope: Americans Reaching for Hope*, Knopf, New York, NY (Reviewed by Sarah Smarsh, *New York Times*, 12 January 2020).

Laville, S. & Pegg, D., 2019, www.theguardian.com/environment/2019/oct/10/fossil-fuel-firms-social-media-fightback-against-climate-action.

Leonhardt, 'G.D.P. is broken, but we can fix it', *New York Times*, 16 December 2019, p. A31.

Levitsky, S. & Ziblatt, D., 2018, *How democracies die*, Penguin, New York, NY.

Lewis, M., 2018, *The fifth risk*, W. W. Norton, New York.

Lipton, E. & Ivory, D., 2017, 'Under Trump, E.P.A. has slowed actions against polluters, and put limits on enforcement officers', 10 December, www.nytimes.com/2017/12/10/us/politics/pollution-epa-regulations.html.

Maciag, M., 2019, 'How many local governments is too many?', *Governing*, 7 May, www.governing.com/topics/politics/gov-most-local-governments-census.html.

Marcuse, P., 2020, 'Sustainability is not enough,' in T. Angotti (ed.), *Transformative planning: Radical alternatives to neoliberal urbanism*, Black Rose Books, Chicago, IL, 104–106.

Marris, E., 2020, 'How to stop freaking out and tackle climate change', *New York Times, Sunday Review*, 10 January, p. 7.

Marshall, M., 1996, 'Germany's social democrats delay Kohl's austerity plan', *Wall Street Journal*, on-line, 22 July.

Martin, I., 2008, *The permanent tax revolt*, Stanford University Press, Stanford.

Martin, I., 2013, *Rich people's movements: Grassroots campaigns to untax the one percent*, Oxford University Press, Oxford.

Martine, G., Alves, J.E. & Cavenaghi, S., 2013, *Urbanization and fertility decline: Cashing in on structural change*, IIED Working Paper, December, https://pubs.iied.org/pdfs/10653IIED.pdf.

Marx, L., 1964, *The machine in the garden: Technology and the pastoral ideal in America*, Oxford University Press, Oxford.

Massey, D. & Denton, N., 1993, *American apartheid: Segregation and the making of the American underclass*, Harvard University Press, Cambridge, MA.

Mayer, J., 2020, 'How Mitch McConnell became trump's enabler-in-chief', *The New Yorker*, 20 April.

McCarthy, N., 2019, 'Oil and gas giants spend millions lobbying to block climate change policies', *Forbes*, 25 March, www.forbes.com/sites/niallmccarthy/2019/03/25/oil-and-gas-giants-spend-millions-lobbying-to-block-climate-change-policies-infographic/.

Mendez, M., 2020, *Climate change from the streets: How conflict and collaboration strengthen the environmental justice movement*, Yale University Press, New Haven.

Miller, R.W., 2017, 'Michael bloomberg: "Nothing Washington can do to stop" action to curb climate change', *USA Today*, 11 November, www.usatoday.com/story/news/world/2017/11/11/michael-bloomberg-jerry-brown-climate-change/854961001/.

Morse, S., Simpson, C.A. & Stein, S., 2020, 'Policing, incarceration and the militarization of urban life', Chapter 7, *Angotti*, pp. 207–226.

NASA, n.d.a., *Is it too late to prevent climate change?*, viewed 15 March 2020, from https://climate.nasa.gov/faq/16/is-it-too-late-to-prevent-climate-change/.

NASA, n.d.b., *The causes of climate change*, viewed 15 March 2020, from https://climate.nasa.gov/causes/.

Nath, J., 2015, viewed 2 April, from www.wesa.fm/post/report-pennsylvania-suffers-devastatingly-large-education-funding-gap#stream/0.

O'Connell, M., 2018, 'Why silicon valley billionaires are prepping for the apocalypse in New Zealand', *The Guardian*, 15 February.

O'Donnell, K., 2020, 'Trump moves to roll back Obama housing desegregation rule', *Politico*, 6 January, www.politico.com/news/2020/01/06/trump-roll-back-obama-housing-desegregation-094874.

Olpadwala, P. & Goldsmith, W.W., 1992, 'The sustainability of privilege: Reflections on the third world city, poverty, and the environment', *World Development* 20, 4 April.

Osnos, E., 2020, 'How Greenwich republicans learned to love Trump', *The New Yorker*, 3 May.

Perry, A.N., Harshbarger, D. & Romer, C., 2020, 'Mapping racial inequity amid COViD-19 underscores policy discriminations against Black Americans', *Brookings*, www.brookings.edu/blog/the-avenue/2020/04/16/mapping-racial-inequity-amid-the-spread-of-covid-19/.

Piketty, T., 2014, *Capital in the twenty-first century*, Harvard University Press, Cambridge, MA.

Piketty, T., Saez, E. & Zucman, G., 2018, 'Distributional national accounts: Methods and estimates for the United States', *Quarterly Journal of Economics* 133(2), 553–609.

Raworth, K., 2017, *Doughnut economics: Seven ways to think like a 21st-century economist*, Chelsea Green Publishing, White River Junction, VT.

Regan, M.D., 2017, 'U.S. cities, states pledge support for climate accord', *PBS News Hour*, www.pbs.org/newshour/nation/u-s-cities-states-pledge-support-for-climate-accord.

Rodden, J., 2019, *Why cities lose: The deep roots of the urban-rural political divide*, Basic Books, New York, NY.

Rothman, J., 2020, 'Same difference: What the idea of equality can do for us, and what it can't', *New Yorker*, 13 January.

Schneider, W., 1992, 'The suburban century begins', *The Atlantic*, July.

Schwartz, J., 2019, 'New York climate lawsuit fails as judge rules for Exxon Mobil', *New York Times*, 11 December.

Starr, P., 2019, 'The battle for the suburbs', *New York Review of Books*, 26 September.

Suarez-Villa, L., 2015, *Corporate power, oligopolies, and the crisis of the state*, SUNY Press, Albany, NY.

Thatcher, M., 2013a, www.theguardian.com/politics/2013/apr/08/margaret-thatcher-quotes 8 April.

Thatcher, M., 2013b, www.theguardian.com/society/2013/apr/09/margaret-thatcher-policies-poor-society.

Torrey, B.B., 2004, 'Urbanization: An environmental force to be reckoned with', *Population Resources Bureau*, 24 April, www.prb.org/urbanization-an-environmental-force-to-be-reckoned-with/.

United Nations, Department of Economic and Social Affairs, 2018, www.un.org/development/desa/en/news/population/2018-revision-of-world-urbanization-prospects.html.

U.S. Energy Information Administration, 2011, viewed 10 March 2020, from www.eia.gov/environment/emissions/ghg_report/ghg_methane.php.

VanHeuvelen, T. & Copas, K., 2019, 'The geography of polarization, 1950 to 2015', *The Russell Sage Foundation Journal of the Social Sciences 2019 September* 5(4), 77–103, www.rsfjournal.org/content/5/4/77.

Warner, S.B., 1962, *Streetcar suburbs: The process of growth in Boston (1870–1900)*, Harvard University Press, Cambridge, MA.

We Are Still In, n.d., www.wearestillin.com/about.

Williams, R., 1973, *The country and the city*, Oxford University Press, Oxford.

Wood, R.C. and Almendinger, V.V., 1961, *1400 governments: The political economy of the New York Metropolitan region*, Harvard University Press, Cambridge, MA.

4 Tent city urbanism[1]

Andrew Heben

Introduction

In February 2009, the American "tent city" suddenly took center stage following a special report on the Oprah Winfrey Show. Through the lens of a sprawling encampment along Sacramento's American River, the report aimed to "humanize" the economic recession and housing foreclosure crisis occurring at the time. Tent cities were reported to be "makeshift shelters set up by people who have lost their homes and have nowhere to go." Each person who appeared on screen gave a heart-wrenching story of recently losing his or her job to the recession, followed by losing his or her home to foreclosure. The conditions in Sacramento were said to be just one example in the "explosion of tent cities across America."[2]

Concurrently, Justin Sullivan of Getty Images released a photo essay documenting the very same camp.[3] Existing conditions were juxtaposed with photographs of similar camps along the American River during the Great Depression – an unmistakable reference to the start of a Second Great Depression. These older camps of the 1930s were known as "Hoovervilles," named after President Herbert Hoover, who was commonly thought to have been responsible for the desperate conditions that led to the development of the sprawling shantytowns in cities throughout the country.

The photographs evoke sympathy by capturing the forlorn expressions of the victims with close-ups of the meager possessions that filled the camp. There is one, however, that stands out from the dismal tone set by the rest. Set in the depression days of 1936 Sacramento, it depicts a middle-aged man leaning back in his lawn chair with laundry hanging on a line in the background. He wears a wide smile as a small child and dog jump on to his lap while another young boy leans over the scene, also smiling. It reveals a rare glimpse into the lighter side of camp life, highlighting the resilience of humans to adapt when presented with challenges. Their possessions may be few but they will endure as a family.

Following this initial coverage, the tent city story was exhausted by news sources across the country. Headlines read "Slumdog USA" and "Cities Deal with Surge in Shantytowns." The stories described the conditions in Sacramento along with similar tent cities that seemed to have abruptly surfaced throughout the country as a result of economic disaster. Today's camps were satirically referred to as "Bushvilles" or "Obamavilles" depending upon the political affiliation of the source.

Why were America's homeless suddenly receiving so much attention when the news media typically avoids intractable social issues at all costs? As one journalist put it, "The story is compelling precisely because it is so visual."[4] Homelessness is usually a piecemeal experience for the housed. Walking in an urban area, one is likely to pass someone holding a sign asking for money, pushing a shopping cart full of personal belongings, or sleeping in a doorway. These are frequent but isolated encounters. The number of people going unhoused may be large but they are dispersed, and the true scale of the issue is hidden. But with a photograph of a tent city, the housed can now place an image to the overwhelming numbers reported by regular point-in-time homeless counts. The experience is no longer isolated and the scale of the issue becomes unavoidably obvious. The same journalist expands on the appeal behind this story:

> The poverty doesn't need to be explained. Instead it can be shown, along with a caption explaining that this is the new face of homelessness and poverty in post-boom, recession-era America. A gleaming city skyline as a backdrop to immiseration. Rio, or Mumbai, on the American River.[5]

The tent city story of 2009 went on to receive coverage from major international news sources like the BBC and Al Jazeera. These reports used the informal camps to illustrate just how bad conditions had gotten in the United States – the world's economic leader now has shantytowns! The UK's Daily Mail ran the headline "The credit crunch tent city which has returned to haunt America," which dramatically described the scene:

> *A century and a half ago it was at the centre of the California gold rush, with hopeful prospectors pitching their tents along the banks of the American River. Today, tents are once again springing up in the city of Sacramento. But this time it is for people with no hope and no prospects. With America's economy in freefall and its housing market in crisis, California's state capital has become home to a tented city for the dispossessed . . . The tents and other makeshift homes have sprung up in the shadow of Sacramento's skyscrapers.[6]*

While these stories may be eye-catching, they are remarkably superficial and lack any real substance other than highlighting the fact that people are, out of necessity, living in tents. The tent cities were ubiquitously portrayed as mere symbols of poverty – the physical manifestations of our nation's housing foreclosure crisis – with no attention given to the longevity, organization, or diversity of the camps. Instead, they imply disorder and strive to evoke awe, sympathy, and sometimes disgust. But did these informal settlements really just appear out of thin air following economic hardship? Even the homeless advocates interviewed in the stories seem to focus only on the tragedy of the situation – that the encampments are a testament to our failure to provide adequate affordable housing.

By March 2010, the National Coalition for the Homeless released a special report – "Tent Cities in America" – that provided more detailed information on the so-called recent phenomenon. "Tent Cities are America's de facto waiting room for affordable and accessible housing," writes executive director, Neil Donovan. "The idea of someone living in a tent in this country says little about the decisions made by those who dwell within and so much more about our nation's inability to adequately respond to our fellow residents in need."[7] Similar to the media's perspective, advocates for the homeless continue to point to the camps as tragic symbols of just how bad conditions have gotten. Tent city is a state of emergency from which people must be rescued, but how and where to is left unclear.

Originally intending to provide national coverage through a series of regional reports, the West Coast was the only region that ended up being surveyed by the Coalition. This was an appropriate starting point since the region contained the most established camps in the country, some of which were among the first to be formalized and regulated. Tent cities are defined as "a variety of temporary housing facilities that often use tents." Though not terribly descriptive, it demonstrates a more investigative approach to the issue and implies an exploration of the various types of camps. The names and locations of select tent cities are listed along with key characteristics of each. "Encampments range in structure, size and formality," the document reports, "Larger more formal tent cities are often named and better known, but don't represent the majority of tent city structures or residents, found with smaller populations and dimensions."[8]

The report gathers ground level research in the field, providing a much-needed analytical perspective of the issue. It becomes evident that while our attention to tent cities may be related to economic conditions, their existence is not. They are in fact a deeply rooted trend in the American city. Examples such as Dignity Village in Portland and Tent City 3 in Seattle were both established in 2000, during times of economic prosperity. Furthermore, while the recent tent city in Sacramento was extensively compared to the Hooverville that existed there during the Great Depression, it is pointed out that camps have existed in some form or another in that area along the American River *since* the 1930s:

> The banks of the American and Sacramento Rivers in downtown Sacramento *have long been a site for homeless encampments dating back to the Great Depression. There have been dozens of scattered campsites for decades along the rivers and in the areas close-by. Periodically law enforcement would dismantle the settlements and take the possessions of many of the homeless people, claiming that they had the legal right to confiscate property under the city's harsh anti-camping ordinance. After a federal civil rights lawsuit was brought against the city and county of Sacramento, an unannounced, informal moratorium on enforcement of the anti-camping ordinances ensued. This allowed the growth of "Tent City," with hundreds of campers congregated on one site because the city and county felt vulnerable to further costly litigation. . . . The American River encampment had been a relatively small settlement until the unannounced moratorium on the anti-camping ban took effect*

after the lawsuit was filed against the city, at which point it quickly grew and stabilized with 100–250 campers at any given time.[9]

While it may be true that this was the largest camp the area had seen since the Hooverville days, the report highlights that the recent economic recession was not to blame for the size as much as the change in the enforcement of laws that make these living conditions illegal. In a survey of 97 Sacramento campers in March 2009, 35% reported that they had become unhoused in the past year.[10] The proportion is certainly high, but it does not represent the majority of the population being served by the tent city. So, while the informal settlements may be influenced by economy, they are affected more so by policy.

Anti-camping ordinances are now commonplace in cities throughout the United States, but Sacramento is home to one of the strictest. The city prohibits camping on private property, even with permission, for longer than 24 hours. Relaxing the enforcement of this law is what allowed the camp to swell. The media frenzy only brought widespread attention to the camp after placing a global spotlight on it, leading embarrassed city officials to disband the camp shortly thereafter on grounds of unsafe and unsanitary conditions.

Defining the American tent city

A tent city is a well-rooted homeless encampment, often with some level of organizational structure and an indefinite population. There is no set size at which an encampment becomes a tent city. Instead, it is an abstract label that is eventually adopted by its inhabitants and sometimes by the surrounding community – providing a sense of identity to an otherwise unrelated group of people experiencing homelessness.

Tent cities can be categorized into two main types based on legal status. Some are sanctioned, meaning they have been formalized and regulated by a municipality in some way or another. All others can be referred to as unsanctioned, meaning they exist illegally by squatting on public or private land. There are a variety of ways in which municipalities have gone about sanctioning a tent city – including conditional use permits, planned unit developments, zoning for camping, and emergency orders. While the "Tent Cities in America" report focuses on these sanctioned examples, comparatively they are rare and the majority of camps are unsanctioned and undocumented. The sanctioned examples are, however, leading a movement that is offering innovative solutions to a situation that is largely regarded as intolerable or deviant.

The designation of sanctioned or unsanctioned status is not always obvious, though. Once discovered, many camps exist in a grey area where they are technically illegal, but there is no foreseeable action taken to evict them. This grey area is the result of the ethical dilemma presented by the issue, and the contentious nature of laws that make such simple acts like sleep illegal. While not sanctioning the tent city, some city leaders unofficially agreed to not evict a specific camp – such was the case with Seattle's Nickelsville, which informally accommodated over 100 unhoused individuals, couples, and families. Others enacted a moratorium on the camping ban – often as a result of lawsuits – that allow camps to informally expand like the example in Sacramento.

While these camps are still considered unsanctioned, they become significantly more stable, which leads to the next method for categorizing tent cities – level of organization. Some are self-organized, developing some sense of order through democratic meetings and community agreements. Others are unorganized, consisting of people simply living next to each other out of necessity. Granted, there is always some level of organization in any human settlement, but these examples consist of a minimal amount. Similar to legal status, many exist somewhere in between these distinctions, demonstrating aspects of both ends of the spectrum.

In the process of sanctioning a tent city, there are diverse models that have emerged for doing so, often influenced by the amount and source of funding. All sanctioned examples are supported by a partnership with a non-profit organization, but the nature of this relationship varies widely. Some organizations support a self-managed model, leaving decision-making, operation, and maintenance in the hands of the residents themselves. These are typically citizen-driven initiatives that require minimal funding. Others adopt a more charitable model, organized similar to the traditional social service model where help is handed down from specialist to client, using public funding sources.

Together, the variable dynamics of legal status, level of organization, and type of funding create a conceptual field on which each tent city can be placed. Where a particular tent city falls in this field often influences the permanence of the camp – certainly one of the most controversial points of discussion around this issue. Are they to be temporary, providing emergency shelter under special circumstance for a limited amount of time? Or are they more permanent, filling a gap in our existing housing system? If so, must they be itinerant, and required to move locations after a set duration of time?

In a paper focusing on the politics of tent cities, Katherine Longley discusses how the permanence of tent cities is always in question. She describes the space in which tent cities exist as a "constant battle-ground" where members of a tent city are striving for a "semblance of permanence from day to day" while the larger community is unwilling to recognize the tent as a "viable permanent dwelling" and instead sees the tent as a "symbol of poverty" that threatens the stability of the surrounding neighborhood.[11] She concludes that this is an unwinnable battle for both sides. While the larger community may succeed in evicting the tent city from a specific site, they will likely fail in removing it from the community altogether. Instead, the tent city will simply move to a new location.

Due to this conflict of interest, choice of location becomes a critical factor in determining the longevity of a tent city. Some are hidden, tucked into a defensible residual space. Forested left-over spaces carved by highways or rivers are often ideal locations for this. Others are more visible, existing within the public's view. These examples are often quickly dismantled unless they can be positioned as a protest with adequate political support. As a result, most unsanctioned camps take to hiding to establish a better sense of stability. If a visible location is sought out, it is typically either an attempt to make a statement or a sign of more lenient law enforcement. Sanctioned tent cities present a similar trend. If a marginalized location is established at the edge of the city, far away from residential zones, the permanence of the community is less likely to be in question than if it takes a central location within the urban core. As a result, a legal place of refuge is often established at the expense of further alienating the population from the rest of the city.

Controlling and reclaiming space

The right to space in the city is ever-changing with the tides of control and resistance. This creates transitional spaces, where the physical space in question is constantly being claimed and shaped by opposing forces. The unhoused – those without the right to a space of their own – must live their entire lives within the public realm, and so there is little distinction between public and private. Cities often confront this uncomfortable reality through the use of environmental design that alters public space to be unaccommodating to those who must live there. Classic examples in public parks include designing benches so that people cannot lie down and special trash cans that restrict gleaning cans and bottles. Downtown exclusion zones go further to prohibit specific activities – often geared toward those not participating in commerce – and ban those with citations from the defined area.

Residual public spaces are defended through other tactics. Some cities have taken to cutting down trees around highway interchanges to expose unwanted inhabitants, and landscaping with large rocks under bridges to eliminate the possibility. Even the selection of certain plant species can be employed to prevent people from sleeping in hidden areas between the plantings and a building.

As a result, the unhoused are often left to develop alternative and innovative techniques for resisting these measures of control. The self-organized tent city is a physical embodiment of this resistance, where the unhoused are reclaiming a space of their own within a community. The American tent city, therefore, is a non-violent revolution that directly responds to the absence of place and participation in today's city. Just as formal actors are trying to control space, those with no place else to go are reclaiming it. As part of their resistance, the unhoused claim autonomy for their cultural and physical place and continually remake it to fit their needs – just as the formal actors continually tear it down to remake it according to their own designs.

Looking at the post-modern American city, we see a persistent homeless population, tightening budgets, and the damaging effects of some people not having a stable space to call home. We see the need for some kind of security in place, a sense of place, of purpose, and of belonging. At the same time we see vacant and underutilized land, an excess of materials going unused, and lots of people who want and need something meaningful to do. It seems only logical then to better utilize the resources available within our local communities to develop sensible, low-cost solutions.

Instead, we rely almost exclusively on federal and state funding to pay service providers that hand help down. The familiar goal to "end homelessness" – both locally and nationally – is almost always sought through plans that attempt to modify people in order to fit a narrow conception of space. Thought is rarely given to the other side of the spectrum – altering our ideas of urban space to fit the needs of people.

Rather than continuing to seek eviction, maybe we should be looking to accommodate these alternative forms of emergency shelter to develop a form of transitional and affordable housing suitable to this population. Currently, there is a clear dichotomy between the formal city solution and the informal human solution. We need cities with formal human solutions. The task requires collaboration and careful compromise to develop a solution that works for all parties involved. The result is a city with a greater sense of social inclusion, which can be beneficial to everyone.

The American understanding of space has been heavily influenced by the grid plan that defines the layout of the American city. Mark Lakeman, co-founder of Portland's City Repair project, often points to this as the root cause of the prevalent social isolation that has left much of today's cities in disrepair. The grid plan was first developed between 3,000 BC and 200 BC in the Euphrates Valley, originating as a military camp plan devised as a means to divide and conquer. It has since become well respected as an efficient and convenient form of city planning. While most US cities are now designed around the grid, few recognize that this was predetermined by federal mandate.[12]

The earliest New England colonies had concentric layouts, radiating out from a central meeting house and town square, and the grid was largely unknown to these modest, self-managed villages. The settlements were also geomorphic in that they adapted to the context established by natural features. The communities were self-contained, both economically and culturally, and so did not accommodate continual growth and expansion. But with the transition from the colonial village to the commercial town at the end of the 17th century, "The common concerns of all the townsfolk took second rank: the privileges of the great landlords and merchants warped the development of the community."[13]

With the enactment of the National Land Ordinance of 1785, the remaining frontier was divided into a continental grid, and all towns and cities established thereafter were required to be developed using a grid plan. However, Lakeman emphasizes that the ordinance lacks any mandate for public squares, and notes that the absence of these places combined with the grid plan had austere consequences on the cities established during westward expansion:

> Only a decade after the public spaces of the commons had been so crucial to the American Revolution, how could the Continental Congress have forgotten such an evidently vital provision as public gathering places? By this omission, the grid of the urban realm in America is reduced virtually to its original form in Assyria. Mitigated by many political freedoms, the provision of parks, the inspiring presence of a dynamic, open landscape, and a recent

history of social and economic improvements through struggle, the American grid nonetheless remains a powerful geometry of isolation which encourages conformity and disassociation . . . The placeless geometry of the grid, which serves to defeat localized social identity and eliminate local community space, always establishes a standardized landscape designed to engender a homogenizing effect on its inhabitants.[14]

Power and control is embedded in the very core of our urban design, which has had significant implications on the way we think – or maybe more appropriately, don't think – about the use of physical space. The grid largely ignores social and natural concerns, and instead reinforces political and economic structures based upon competition and expansion above all else.

The consequences are particularly harsh for those left without a right to a piece of the grid, which purposefully keeps property and economic freedom out of common reach. Where are they supposed to go? Naturally, this population has historically established off-the-grid alternatives to simply get by from day-to-day, as exemplified by the self-organized tent cities covered in this book. However, over the course of the past century, policies have been put in place to further tighten the control of our physical environment. What has ensued is similar to the once popular "whack-a-mole" arcade game. Just as policy makers cover one hole in the grid, the same issue persistently pops out of a new one.

The demise of low-cost housing

In the past, the grid city was also mitigated by a diversity of low-cost housing options in the private market. An abundance of single room occupancy (SRO) hotels flourished in US cities of the early 20th century, ranging from rooming houses for the middle class to lodging houses for the lower class. Just within lodging houses, accommodations ranged from private rooms (dorms) to large rooms broken into cubicles (cages) to bunk rooms with rows of beds to a few square feet of space on a floor (flops) – with each getting progressively cheaper. As a result the homeless and nearly homeless had a variety of alternative forms of shelter to choose from, which has been well described by Hoch and Slayton:

In many cases the choice was made for the individual, in a sense, by economic conditions. Housing was a fluid and uncertain situation; eventually most of the population used all the alternatives – cages, flops, dorms, and the streets – at one time or another. When this law did not apply, when a very minimal amount of discretionary income was available, the hobo or tramp or bum was a fee agent. And whenever this was the case, the housing of choice was the cage hotel . . . The reason for the success of the cage hotels, aside from the fact they filled the desperate need for housing, was that they provided other advantages: privacy and freedom. Every room had its own lock and key, and the resident could come and go as he pleased. . . . The popularity of cage hotels did not stem from any great improvement in the physical setting (which was marginal at best), but rather in the qualitative benefits of a private room.[15]

Essentially, classic SRO housing consisted of very compact private spaces – which itself varied significantly based upon price – along with common facilities that usually included a bathroom down the hall and a shared kitchen and dining room on a separate floor. The dense housing was also supported by mixed-use neighborhoods where surrounding stores and public spaces became extensions of each resident's home. This arrangement was valued not because it met certain material standards, but because it was economical yet still respected an individual's privacy and autonomy.

It is important to note that the availability of this housing option did not necessarily result in a concentration of those on the bottom rung of society. The arrangement also appealed to a healthy mix of the middle class – including young adults looking to start a career, artists, students, and blue-collar workers. It was not until suburbanization, the accompanying ethos of individualism, and eventually the deinstitutionalization of mental health facilities that led to a homogenized population of tenants who were down-and-out. In addition, SROs were crippled by urban policies intended to eliminate traditional low-income housing, a changing economy with diminishing demand for transient labor and higher rates of unemployment, and welfare plans that undermined the autonomy of the poor.[16]

The vibrant, walkable neighborhoods decayed as urban areas experienced widespread disinvestment. Attention was instead turned to auto-oriented development just outside the city. And by 1950, more people would live in suburbs than in cities. Many jobs were still downtown, but highways, parking garages, and skyways allowed for a commute without ever stepping foot on a sidewalk. Beginning with the "urban renewal" programs of the 1960s, neighborhoods identified as a blight were completely razed – often as a result of conveniently routed highway construction. The existing SRO districts were depleted, and often replaced with a smaller quantity of single-use apartment buildings.

Furthermore, zoning and building code provisions had been put in place to ensure the demise of SRO type housing. Zoning was used to corner the use in the oldest parts of town, and changes to the building code made it difficult, if not impossible, to build new SRO housing.

While housing reformers – both well-intended and self-seeking – celebrated improvements in the minimum standard of living, the act removed the bottom-end of private housing from the market by making it too expensive to develop and manage. This has led to the current dilemma where, in order to be affordable to low-income tenants, housing is inherently dependent on government subsidies that have been made to be in short supply. Today, the construction of low-income housing comes at a cost of upwards of $200,000 per unit.

In his short e-book on this subject, Alan Durning argues that it is once again time to allow the generation of inexpensive, housing options to supplement the shortfall of public subsidies. And, he points out a way for doing so that, while vehemently guarded, is not as complex as we might think – we just need to do away with regulations that simply protect adjacent property values by defining "decent housing" based on middle-class expectations.[17]

The situation highlights the influence of the grid, which reinforces attention to property value where the house is a commodity to be bought and sold. This has created an inflated standard of living where middle-class values determine what is and isn't acceptable – even when some are content with less. While limiting profit-hungry landlords from taking advantage of tenants is a legitimate concern, minimum square footage, kitchen, bathroom, and parking requirements go beyond the building codes foremost focus of life safety provisions and structural collapse. Instead, under the shroud of health and welfare, the code has come to mandate middle-class norms and eliminate simpler housing options that are perceived to negatively influence adjacent property value. But by failing to ensure that everyone can meet those standards, we have in fact jeopardized the life safety of an entire segment of the population.

Broken windows in the new urban frontier

By the 1980s, space in some city centers began to receive a renewed sense of value. Urban areas became a new "wild west" for artists followed by young professionals, creating what Neil Smith called "the new urban frontier."[18] By the time development had finally sprawled as far as it could functionally reach from the city, the urban center once again became a place to explore and tame. This meant that the cost of renting the most affordable apartments went up, and neighborhoods were gentrified as existing tenants were pushed out. With a renewed interest in urban living, many cities are now reinvesting in what's left of their deteriorating SRO housing stock. However, they are being reincarnated as boutique hotels to spur economic development rather than provide low-income housing.[19]

The Lower East Side of New York City provides an early example of this national phenomenon as described by Smith. After widespread gentrification of the neighborhood resulted in the infamous 1988 riot at Tompkins Square Park, a few dozen unhoused people began to use the public space as a new home. Many had recently lost their housing nearby but refused to leave the neighborhood, and one resident described it as "the place for one last metaphorical stand" against gentrification. By 1991, there were around one hundred shanties and tents occupying the park when the police evicted over 300 people and quickly barricaded the site with a fence. After closing the park, several smaller tent cities formed in adjacent vacant lots, only to soon be bulldozed and also fenced. Finally, the economic refugees were routed underneath nearby bridges or any available space concealed from public view.[20]

The taming of this new urban frontier, among other things, meant clearing the unhoused from sight. As cities became fixed on measures of "livability" in an effort to attract capital, local governments began to practice "the annihilation of space by law" – using legal remedies to physically erase the spaces in which the unhoused must live.[21]

Kelling and Wilson justified this initiative in <u>1982</u> with their "broken windows" theory, which offered a new approach for policing urban areas. The theory asserts that a broken window in a city implies that "no one cares," and will inevitably lead to more broken windows. An experiment of two untended cars – one in the Bronx and one in Palo Alto – was used to test the theory. Both cars were eventually heavily vandalized, though the more prominent location in Palo Alto needed a little kick-start before the defacement ensued. From here, Kelling and Wilson make the conclusion that social ills that are visible in a city will inevitably lead to more, and likely worse, social decay. They assert, "the unchecked panhandler is in fact the first broken window."[22] In other words, broken people in public spaces results in more broken people.

The theory suggests a strict enforcement of "quality of life crimes" such as panhandling, loitering, and public intoxication or urination – with the idea that these acts infringe on the quality of life for others in the city. The phrase is quite ironic since many of these crimes are actually a consequence of the quality of life of the offender. In the past, these minor acts had been commonly overlooked in urban areas in order to concentrate policing efforts on more severe crimes. However, the broken windows theory asserts that it is in fact these minor crimes that lead criminals to believe they can get away with much more serious crimes like rape, murder, and theft. So by eliminating petty crimes, a city could in turn prevent more violent crimes. The widespread acceptance of this premise marks a drastic shift from "servicing" toward "policing" as a method for managing the issue of homelessness.

With the rise of Mayor Rudolph Giuliani in 1994, New York City became the first city to fully adopt the broken windows theory. The so-called quality of life crimes were enforced more strictly than ever, and the unhoused were either arrested or told to move along. Giuliani pro-claimed:

> You are not allowed to live on a street in a civilized city. It is not good for you; it is not good for us. Maybe the city was stupid enough to embrace that idea years ago. We care about you enough and are not that dumb to think you should live on streets. So we contact and tell you, you have got to move.[23]

As is common, the only problem with the mayor's benevolence was that he did not provide a viable alternative to those living on the street. In fact, his first budget proposed that the homeless pay rent for nights spent in city-run shelters.

New York City's focus on petty crimes did see promising results. Crime rates dropped, and in 2002 Manhattan had the lowest murder rate it had seen in a century. The number of homicides in 2000 dropped 73% when compared to 1990.[24] But did citing panhandlers really deter murder? It is not evident how much the enforcement of quality of life crimes contributed to the shift. Experts have offered everything from economic prosperity to legalized abortion as a reason for the significant reduction in crime. Levitt and Dubner argue that removing "broken windows" actually had little effect at all. They cite two reasons to support this claim. First, the homicide rate had already dropped 20% before Giuliani was elected – meaning the shift was well under way prior to the new policing strategies being adopted. Second, the New York Police Department was expanded by 45% between 1991 and 2001 – over three times the national average during this period.[25] A more logical reason for the reduction in crime, then, may have been the drastic increase in police force, not necessarily the change in strategy. Regardless, after New York City's success, cities across the country began to institute the broken windows theory, cracking down on quality of life crimes.

While conversations around policing the homeless are often varied and distorted by political ideology, evaluating the practice from an economic perspective is difficult to dispute. What is the cost of controlling space? While some may agree with these policies, do they want to personally pay for it with their tax dollars? A study comparing nine major US cities found that incarceration and hospitalization are far more expensive routes than providing shelters or even supportive housing. Of the nine cities surveyed, the average cost per day for each setting was: hospital, $1,638; mental hospital, $550; jail, $81; prison, $79; supportive housing, $30; shelter, $28.[26] Simply providing a legal "place to be" would therefore benefit both the unhoused and the fiscal conservative. A recent study in Central Florida quantifies this economic benefit – showing that the public expenses that are incurred by leaving someone on the street comes at an annual cost of $31,065 per person. Meanwhile, providing that person with permanent housing, job training, and health care was found to cost taxpayers 68% less at $10,051.[27]

Reclaiming democracy

While the conflict has been presented as a battle for space, it is usually a rather one-sided fight with formal actors wielding all the power, and the informal actors doing the best they can to stay out of sight. But in the past, the tent cities and shantytowns of America have located in more public spaces. For example, during the early 1930s, there was an infamous "Hooverville" located on what is now the Grand Lawn of New York City's Central Park. Recognizing the severity of the economic situation and the widespread unemployment of the time, it was recorded that the public and political sentiment was largely with the shantytowns. Contrary to today's coverage, they were described as symbols of "ingenuity." In 1932, The New York Times reported that the city's Park Department halfheartedly intended to bulldoze the shantytown. "We don't want to do it but we can't help it," said the Deputy Parks Commissioner, adding, "although the men had maintained good order, had built comfortable shacks and furnished them as commodiously as they could, there were no water or sanitary facilities near the settlement."[28] The Tompkins Square Park camp described earlier is another very public example that was around for years in the 1980s. However, following the adoption of the broken windows theory, visible public space ceased to be a viable option for most.

In their article "Reclaiming Space," Groth and Corijn advocate for informal actors – such as the members of tent cities – to take a stronger role in setting the urban agenda. They begin with the assertion that post-modern planning has failed to respond to post-modern urbanism. Comparatively, the modern city consisted of a homogenized population with a clear agenda, and centralized planning agencies were able to use order and rationality to serve this relatively uniform society. The result was what Jane Jacobs described in 1961 as "a city of monotony, sterility, and vulgarity."[29]

On the contrary, the post-modern city has witnessed the emergence of a more pluralistic society with highly differentiated agendas. As a result, post-modernist planning theory has attempted to become more flexible, but Groth and Corijn argue that planners still encourage spaces "catered for a relatively uniform society in a system of mass production and mass consumption."[30] Formal strategies limit the complexity of the city and aim to establish a predictable population. This diminishes the "dimension of socioeconomic richness and cultural mobility upon which the traditional metropolis thrives," and instead we are left with a streamlined city, where "staged images of the public replace the spaces of idiosyncratic interaction."[31]

"Reclaiming Space" calls for an alternative approach in which urban spatial structures diverge from "active repossessions" and "symbolic reconstructions" to create a city with a greater sense of social inclusion. Residual space, which lacks any significant economic value, is identified as a suitable place for this type of urban transformation to occur. Here, Groth and Corijn argue that formal planning and politics should step aside to allow space for more informal development.

Today's tent cities organized by the unhoused offer a prime example of this type of reclamation of space. Due to the negligence of formal actors, marginalized members of the city are taking matters into their own hands. Out of necessity, they are rediscovering the power of community, and through this collective effort, people in a similar situation are forging their own solutions by claiming space and working together to improve their individual situations. Rather than settling in popular public spaces, these economic refugees tend to seek hidden residual spaces where they are far less likely to be bothered. Camps tend to most often form in the left-over spaces carved by highways, railroads, or rivers. The latter has proven problematic for many cities, with the unhoused locating in environmentally sensitive areas like wetlands and floodplains since these are often large tracts of land isolated from the public. While they are out of sight, the lack of infrastructure can lead to the degradation of these protected ecological areas. This is probably the most reasonable justification for evicting a camp – and tends to receive public support from both the right and left – but it is typically out of a lack of options that these sites are chosen. This makes an environmental case for establishing a legal "place to be" on a site that is appropriate for human habitation.

But the endorsement of the broken windows theory suggests that these informal settlements are robbing a certain degree of quality of life from the surrounding, housed community. Furthermore, the settlements thwart the efforts by formal design to establish predictable behavior. As a result, laws and strategies have been adopted to disrupt these acts of necessity, exiling those without a right to space in the city to an itinerant lifestyle.

However, these informal tent communities actually return some quality of life to their inhabitants, and by applying the same logic behind the broken windows theory positively, this benefits the surrounding population as well. Simply allowing for a legal place to reside – even with the most meager provisions of shelter – reduces negative, external impacts on the city. In fact, one could argue that these people are no longer homeless.

Through the practice of direct democracy, these camps can create safe, nonviolent environments with some degree of privacy. Many have established some form of self-governance through regular community meetings where the group adopts basic agreements that, at a minimum, prohibit acts of violence, theft, and illegal drugs within the camp. Those who break the camp's rules are often told to leave, either temporarily or permanently, depending on the offense. The offender typically then has a chance to state his or her case, with a majority vote by the other members determining the outcome.

Because those who break the community agreements are frequently voted out of the camp, these places do not comprehensively address the multi-faceted issue of homelessness. Those with severe addictions and mental health issues will not last long in this kind of organized environment, and other solutions remain necessary for this sector of the chronically homeless. However, the acceptance of organized camps could significantly reduce this population in the future. Not knowing where you will sleep on any given night is enough to drive nearly anyone mad or to resort to vices, and having a secure place to be from the beginning can make all the difference.

These dynamics are shocking to many who presume that the homeless either can't or don't want to organize functionally. But this perception fails to account for the larger theme that unites these places. They often possess an inherent sense of egalitarianism – the belief that we are all in this together – that has largely been forgotten by a society focused on self-interest and personal independence. This quality is partly because each person has an equal amount to lose if the camp is evicted, and so everyone does their best to maintain a presentable environment.

This intimate practice of democracy is eerily absent elsewhere in the United States, yet most citizens still perceive themselves as members of a democracy. However, the term "democracy" was intentionally left out of both the US Constitution and Declaration of Independence, as the framers found the system, in its true form, impractical – and it's clear that this concern has consistently endured since then. David Graeber writes, "As the history of the past movements all make clear, nothing terrifies those running the U.S. more than the danger of democracy breaking out." He therefore concludes, "If we are to live in any sort of genuinely democratic society, we're going to have to start from scratch."[32] And this is exactly what is taking place in today's tent cities that are being self-organized by the unhoused.

The occupy influence

Tent cities once again entered the public arena when the Occupy Wall Street movement swept the country at the end of 2011. The movement emerged as an effort to raise awareness of the societal consequences of extreme economic inequality, influence the existing federal policies that encouraged this disparity, and build solidarity behind a common cause. In a country that is so commonly influenced by money, Occupy attempted to turn the tables by taking advantage of strength in numbers – pitting 99% of Americans against the top 1%, and exposing the contradictions of an elite-run "democracy."

Occupy was physically and symbolically embodied in the form of robust protest camps that reclaimed central public spaces in cities throughout the country. Activists descended upon these spaces, packing them with tents and clever signs that illustrated their cause. In addition to the individual tents, makeshift facilities were pieced together to support living full-time at the camp – including kitchens, clinics, libraries, and a variety of other community-based services. While the tent cities popped up as a demonstration against social and economic inequality, they soon evolved into places of experiment in alternative living. The camps put their political message into practice by operating communities based on direct democracy and horizontal organization. Regular general assembly meetings used consensus decision-making, and they became places where anyone could be sheltered, fed, and heard.

In cities throughout the country, many of the protestors who filled the Occupy camps had experiences that would unintentionally alter the focus of the movement. These places of protest and experiment soon also became places of refuge – providing a stable place to be for the city's unhoused. The camps were strikingly similar to those already being organized by the unhoused elsewhere, yet these examples had been largely planned by otherwise housed protestors. For a brief moment in time the housed and the unhoused were living side by side, sharing meals and making decisions together. And as winter set in, inexperienced campers received survival tips from those accustomed to the living situation.[33]

But by the end of December, local authorities had dismantled most camps in some way or another. Cities had been bending their anti-camping ordinances to allow for the protests – directed at larger scale government – but eventually enough was enough, and the great experiment was laid to rest once again. In Portland, Oregon the park was cleared of the hundreds of tents that had filled it for over a month, and a fence was erected so that no one could return. A plan to occupy a different space shortly after the eviction was immediately shutdown by the police. Without a physical place the movement lost traction, and with time, the expansive scope of the protest seemed to dissolve back into the various factions that originally composed it.

In the wake of the Occupy camp, the initial focus on federal and global policies shifted to local issues, with many groups coalescing around the issue of homelessness after realizing they had a tangible crisis right in front of them in need of attention. Camp life had deeply impacted several otherwise housed activists, creating a passion for putting this alternative living model back into practice – if not for themselves, at least for their previous neighbors who were now left to fend for themselves. As a result, I would argue that the lessons learned internally in these camps were even more influential than the larger points they voiced externally. Some local Occupy movements – most notably those in Eugene, Oregon and Madison, Wisconsin – have since resurfaced as citizen-driven initiatives for establishing legal "places to be" for the unhoused, adding fuel to an already burning fire.

An enlightened vision

By concentrating power in the hands of formal actors, we have not only created a culture of resistance but also a culture of dependence. This depiction is commonly used to stigmatize the homeless, but in reality it is their only legal option. Being independent and developing alternative solutions while being unhoused – such as forming a democratic tent community – often results in citations and arrest, making it even more difficult to ever get a job or housing. Instead, the unhoused are forced into top-down social service programs that tend to initiate this culture of dependence – sending the message: you no longer need to know how to take care of yourself in the absence of resources, but at the same time, we don't have the capacity to help you forever. The unhoused are put on a one-way track where they are told: get in line so that you can pay subsidized rent, and not learn how to build your own small home so you don't have to always pay rent.

This message is not new. It has been around since the beginning of American history, dating back to early westward expansion. The white settlers took the land, and in exchange, the indigenous were confined to certain reservations where they could continue to manage the land independently. But with time, the reservations were continually reduced in size as the thirst for expansion continued. With the Dawes Act of 1887, reservations were required to be divided into saleable parcels, which inevitably disrupted indigenous tribal culture by establishing divided and competing interests. Eventually it reached a tipping point – the reservations became so small that there were not enough animals to hunt for food, clothing, or shelter. Cultures that could once exist independently off the land were cut off from their natural provider, and made dependent on commerce. To mitigate this, the white settlers promised to provide for their basic needs in exchange for the land, but never followed through to the extent they claimed. Here, Lakeman's assertion in his history of the grid rings true – cultures that pre-exist the arrival of the grid tend to disappear.

In the late 19th century, General George Custer led a charge to exterminate the Lakota Sioux, resulting in the infamous Battle of Wounded Knee. While at the time this was seen by the public as a necessary means for westward colonial expansion, by the end of the 20th century, Neil Smith argues, "most of us would have to come down on the side of the Sioux." He draws a comparison between the Sioux and today's homeless who "suffer a symbolic extermination and erasure that may leave them alive but struggling on a daily basis to create a life with any quality at all."[34] Smith then makes reference to the Homesteading Act of 1862, which granted land rights to western pioneers just a few years prior to the massacre of the Sioux. He depicts the early pioneer – commonly portrayed with romantic images of rugged individualism and patriotism – in a much different light:

> *the majority of heroic pioneers were actually illegal squatters who were democratizing land for themselves. They took the land they needed to make a living, and they organized clubs to defend their land claims against land speculators and land grabbers, established basic welfare circles, and encouraged other squatters to settle because strength lay in numbers.*[35]

The key to these early squatters receiving political power was their organization, similar to the tent cities today that have achieved sanctioned status. Smith concludes, "It is just possible that in a future world we may also come to recognize today's squatters are the ones with a more enlightened vision about the urban frontier."[36]

Notes

1 Adapted from chapters 1 and 2 of Tent City Urbanism, self-published by author in 2014.
2 Harpo Productions, "Inside a tent city: a Lisa Ling exclusive," In *The Oprah Winfrey Show*, February 25, 2009.
3 Justin Sullivan, 'From boom times to tent city," *MSNBC*, 2009, www.msnbc.msn.com/id/29528182/displaymode/1107/s/2/
4 Sasha Abramsky, "Tent cities don't tell poverty's full story," *The Guardian*, April 3, 2009.
5 Ibid.
6 Paul Thompson, 'The credit crunch tent city which has returned to haunt America', *Daily Mail*, March 6, 2009.
7 National Coalition for the Homeless, *Tent cities in America: Pacific coast report*, 6, 2010, nationalhomeless.org/publications/Tent%20Cities%20Report%20FINAL%20 3-4-10.pdf.
8 Ibid., pp. 8–9.
9 Ibid, p. 36.
10 Ibid, p. 39.
11 Katherine A. Longley, 'Governing homelessness: The politics of tent cities in the U.S.', *The John W. McCormack Graduate School of Policy Studies* 2–4 (2006).
12 Mark Lakeman, 'A concise history of the grid', 1998.
13 Lewis Mumford, *Sticks & Stones: A study of American architecture and civilization* (New York: Dover, 1955), 35–37.
14 Lakeman, 'Concise history of the grid', 14–17.
15 Charles Hoch and Robert Slayton, *New homeless and old* (Philadelphia: Temple University Press, 1989), 52.
16 Ibid, 62.
17 Alan Durning, *Unlocking home: Three keys to affordable communities* (Seattle: Sightline Institute, 2013).
18 Neil Smith, *The new urban frontier: Gentrification and the revanchist city* (New York: Routledge Press, 1996), 3–27.
19 For example see: "Downtown Cincinnati boutique hotel might replace low-income flats" Cincinnati Business Courier, September 28, 2009.
20 Smith, *New Urban Frontier*, 9.

21 Don Mitchell, *The Right to the City: social justice and the fight for public space*, (New York: Guilford Press, 2003), 161–190.
22 George L. Kelling and James Q. Wilson, 'Broken Windows: The police and neighborhood safety', *The Atlantic*, March 1, 1982.
23 Eric Lipton, "Computers to track quality of life crime, Giuliani says," *The New York Times*, November 15, 2000.
24 Steven D. Levitt and Stephen J. Dubner, *Freakonomics: a rogue economist explores the hidden side of everything*, (New York: William Morrow, 2005), 129.
25 Ibid, 129.
26 Lewin Group, "The Costs of Serving Homeless Individuals in Nine Cities," (2004).
27 Central Florida Commission on Homelessness, "The Cost of Long- Term Homelessness in Central Florida," (2014).
28 "25 IN PARK SHANTIES POLITELY ARRESTED; 'Hoover Valley' Colony in Old Reservoir Raided as Hamlet Is Deemed Health Hazard. ONE SHACK WAS 'RADIO CITY' There Jobless Squatters Gathered for Broadcasts – Other Homes Showed Ingenuity of Builders," *The New York Times*, September 22, 1932.
29 Jane Jacobs, *The Death and Life of Great American Cities*, (New York: Random House, 1961), 7.
30 John Groth and Eric Corjin, "Reclaiming urbanity: indeterminate spaces, informal actors and urban agenda setting,"*Urban Studies*, 42, no. 3 (2005): 504.
31 Ibid, 505.
32 David Graeber, "Occupy Wall Street's anarchist roots," *Aljazeera*, November 13, 2011.
33 Erika Niedowski, "Occupy Wall Street Protesters Prepare for Winter Weather," *The Huffington Post*, October 29, 2011.
34 Smith, *New Urban Frontier*, 230–231.
35 Ibid, 231.
36 Ibid, 232.

References

Abramsky, S., 2009, 'Tent cities don't tell poverty's full story', *The Guardian*, 3 April.
Central Florida Commission on Homelessness, 2014, *The cost of long-term homelessness in Central Florida*, Central Florida Commission on Homelessness, Orlando, FL.
Cincinnati Business Courier, 2009, 'Downtown Cincinnati boutique hotel might replace low-income flats', *Cincinnati Business Courier*, 28 September.
Durning, A., 2013, *Unlocking home: Three keys to affordable communities*, Sightline Institute, Seattle.
Graeber, D., 2011, 'Occupy Wall Street's anarchist roots', *Aljazeera*, 13 November.
Groth, J.,& Corjin, E., 2005, 'Reclaiming urbanity: Indeterminate spaces, informal actors and urban agenda setting', *Urban Studies* 42(3).
Harpo Productions, 2009, 'Inside a tent city: A Lisa Ling exclusive', in *The Oprah Winfrey Show*, 25 February.
Hoch, C. & Slayton, R., 1989, *New homeless and old*, Temple University Press, Philadelphia.
Jacobs, J., 1961, *The death and life of Great American cities*, Random House, New York, NY.
Kelling, G.L. & Wilson, J.Q., 1982, 'Broken windows: The police and neighborhood safety', *The Atlantic*, 1 March.
Lakeman, M., 1998, *A concise history of the grid*, Communitecture, Portland, OR.
Levitt, S.D., & Dubner, S.J., 2005, *Freakonomics: A rogue economist explores the hidden side of everything*, William Morrow, New York, NY.

Lewin Group, 2004, *The costs of serving homeless individuals in nine cities*, Lewin Group, Falls Church, VA.

Lipton, E., 2000, 'Computers to track quality of life crime, Giuliani says', *The New York Times*, 15 November.

Longley, K.A. 2006, 'Governing homelessness: The politics of tent cities in the U.S.', *The John W. McCormack Graduate School of Policy Studies* 2–4.

Mitchell, D., 2013, *The right to the city: Social justice and the fight for public space*, Guilford Press, New York, NY.

Mumford, L., 1955, *Sticks & stones: A study of American architecture and civilization*, pp. 35–37, Dover, New York, NY.

National Coalition for the Homeless, 2010, *Tent cities in America: Pacific coast report*, 6, nationalhomeless.org/publications/Tent%20Cities%20Report%20FINAL%203-4-10. pdf.

New York Times, 1932, '25 in park shanties politely arrested; 'Hoover Valley' colony in old reservoir raided as Hamlet is deemed health hazard, one shack was 'radio city' there jobless squatters gathered for broadcasts – other homes showed ingenuity of builders', *The New York Times*, 22 September.

Niedowski, E., 2011, 'Occupy Wall Street protesters prepare for winter weather', *The Huffington Post*, 29 October.

Smith, N., 1996, *The new urban frontier: Gentrification and the revanchist city*, Routledge, New York, NY.

Sullivan, J., 2009, 'From boom times to tent city', *MSNBC*, www.msnbc.msn.com/ id/29528182/displaymode/1107/s/2/.

Thompson, P., 2009, 'The credit crunch tent city which has returned to haunt America', *Daily Mail*, 6 March.

5 Transition design as a strategy for addressing urban wicked problems

Gideon Kossoff and Terry Irwin

Introduction

An increasingly high proportion of the world's population now lives in cities (Barber 2013; Rockefeller Foundation 2019; Glaeser 2011) which are confronting a wide range of challenges known as wicked problems (Irwin 2011a, 2011b; Australian Public Service Commission 2007; Coyne 2005; Buchanan 1992; Rittel & Webber 1973). Problems such as climate change, drought, obesity, poverty, racism, waste management, homelessness, crime and loss of biodiversity are *systems problems* that require a systemic response (Rockefeller Foundation 2019).

At the level of the city, problems like these become even more complex because they are interrelated, interdependent and affect increasing numbers of people. Conversely, cities are "places of innovation" (ibid, p. 7) and are well positioned to ignite systems-level change by disseminating and sharing innovations with one another. Civic and governmental leaders are often highly motivated to direct significant amounts of social, economic and political power towards wicked problem resolution (Barber 2013; Katz & Nowak 2017) because they suffer the effects of wicked problems directly. This chapter will argue that because these complex, wicked problems cannot be resolved by a single group, department or area of expertise, radical collaboration among diverse sectors *and* the stakeholders affected by the problem will be necessary to resolve them.

Transition design is an emerging, design-led approach[1] for addressing these complex, wicked problems and seeding and catalyzing societal transitions toward more sustainable and desirable long-term futures.

This chapter argues that transition design can be a useful, transdisciplinary approach for the developing *ecologies of solutions* at multiple levels of scale (the region, the city, the neighborhood and the household) (Kossoff 2011, 2019a). Transition design is an approach that could be used by municipalities, city-based organizations (in business, health care, education and the arts) and grassroots community groups to collectively and collaboratively address the wicked problems that they face.

Characteristics of wicked problems

In the latter half of the 20th century, urban planner Horst Rittel coined the term "wicked problem" to describe a class of problem that was deemed "unsolvable" because of its high degree of complexity. He contrasted them with "tame" problems that he argued could be solved using traditional, linear processes based upon predicted outcomes (Rittel & Webber 1973).

About the same time, discoveries in the emerging field of chaos and complexity sciences shed light on both the structure and behavior of complex open systems and identified key principles such as emergent properties,[2] self-organization[3] and sensitivity to initial conditions,[4] (Goodwin 2002; Briggs & Peat 1989; Capra & Luisi 2016; Wheatley 2006) that are also useful in understanding wicked problems and their systems contexts (Irwin 2011a, 2011b). Many of these principles can be observed and explain the way wicked problems manifest at the level of the city.

Wicked problems are multi-scalar, multi-causal and interdependent

Urban problems such as crime, poverty, homelessness and many others, manifest at multiple levels of scale (neighborhoods, districts, the city and even the region), have multiple root causes and are linked to other wicked problems. For example, homelessness in many cities can be directly linked to other wicked problems such as the financial crisis of 2008, opioid addiction, gentrification, racial profiling and a lack of access to affordable education. Issues such as racial profiling can arise from within the city and its neighborhoods, but others like the financial crisis or the national opioid epidemic originate at the national or global level, then cascade down systems levels. These multi-scalar dynamics, which unfold over short, mid and long horizons of time add exponentially to a wicked problems' level of complexity.

Wicked problems have self-organizing features, display emergent properties and are permeated by complex social dynamics

Wicked problems are comprised of designed artifacts, communications, built infrastructure, technology, scripted interactions and most importantly – people – which makes both their structure and behavior highly complex. Chaos and complexity theories show that social systems are highly self-organizing, and their response to perturbations and disruptions from their external environment is self-directed and unpredictable (Wheatley 2006; Byrne 1998). These responses can result in new, unexpected and "emergent" forms of behavior. To extend the example of homelessness within a city, an external perturbation to a homeless population might take the form of police evicting homeless people from parks and other public spaces. In cities like San Francisco, a perturbation like this has resulted in homeless populations self-organizing to form "emergent" settlements in gentrified neighborhoods or along public thoroughfares and streets, which has given rise to a host of new problems related to public health and safety (Trinko 2020).

Wicked problems comprise multiple stakeholders with conflicting agendas and concerns

In this chapter we refer to stakeholders as any group affected by a problem (stakeholders also include other species or members of an ecosystem affected by the problem) (Seed & Macy 2007; Driscoll & Starik 2004). Stakeholder groups often have conflicting definitions of the problem as well as ideas about how to solve it. This is due to differences in socio-economic- political status (uneven power dynamics and access to resources) as well as differing beliefs, values, assumptions, expectations and needs (Andersen & Nielsen 2009; Hiemstra, Brouwer & Van Vugt 2012). For example, in an era of climate change, many cities face serious water shortages (Majumder 2015) which will affect every member of its population. Members of that population, however, fall into myriad, diverse stakeholder groups and most will belong to several different groups simultaneously. The relations among the groups are complex and mean that their ideas about the problem and how to solve it can be in conflict and alignment in relation to the problem's many facets.

Business people who rely upon tourists who come to enjoy local, water-related sports will view the water shortage very differently than residents whose water is rationed; tourists stay for a brief period of time but are able to use as much water as they like, and may even use a city's potable water supply for sports activities. Local farmers who grow water-intensive crops such as avo-cados will require vast amounts of water from local water sources. Both the farmers and business people may view their water use as justified because they employ local people and contribute to the local economy. However, the farmers may resent the tourists and feel there would be no water shortage if tourism was eliminated. A lower-income resident who is surviving on a minimum wage is not only adversely affected by water rationing, but also by the increasing cost of water. This "web" of complex relations among multiple stakeholders often goes unseen (and therefore unaddressed by traditional problem-solving approaches) and yet is a barrier to problem resolution (Irwin & Kossoff 2017a, 2017b; Hamilton 2019).

Wicked problems are governed by feedback loops

Positive feedback loops

Within urban problems, positive feedback loops are the most easily seen because they tend to *exacerbate* wicked urban problems. Positive feedback is present when a small change or intervention in the system rapidly amplifies. The effects of feedback can be growth, chaotic behavior, disequilibrium and overall system instability (Briggs & Peat 1989; Goodwin 2002; Capra & Luisi 2016). For example, a person or family in poverty is often surviving on minimum wages and can easily fall deeper and deeper into debt, because their wage increases do not keep pace with the rising costs of living (rent, childcare, utilities, groceries, etc.). The individual(s) must therefore work longer hours to earn extra money or take on a second job, which prevents them from acquiring additional credentials or schooling that might qualify them for a better paying job. If an illness or other unexpected expense arises, they may fall deeper into debt and will be unlikely to obtain a loan because they have poor credit due to their impoverished circumstances;

the set of circumstances builds upon and exacerbates each other, accruing and worsening the situation.

Negative feedback loops

Negative feedback loops are more difficult to identify but they are often *barriers* to wicked problem resolution. Initiatives aimed at resolving a wicked problem may be launched, but something (negative feedback) prevents the positive change from taking hold. A negative feedback loop often involves attitudes, beliefs, values, behaviors and practices which can represent entrenched ways of thinking and acting. Take as an example a city-wide initiative to conserve water or electricity. The city might provide financial incentives with conservation targets, which, if met, could create sweeping positive change. Despite what might seem to be an obvious and immediate motivator, the initiative might fail because it would require radical changes in users' behavior or practices. Meeting water conservation targets might involve showering for shorter periods of time, watering gardens or washing cars less frequently or even capturing grey water[5] for reuse. The habits of a lifetime cannot be broken via a single financial incentive, so the potential for change is "damped down" by the feedback loop of entrenched, non-sustainable behavior and practices (Kuijer 2014; Shove & Walker 2010; Shove, Pantzar & Watson 2012).

Wicked problems straddle institutional, disciplinary and sectoral boundaries

Urban wicked problems manifest at multiple levels of scale, connect to myriad other wicked problems and straddle institutional, disciplinary and even sectoral boundaries. However, traditional problem-solving approaches tend to frame and address problems within disciplinary silos and/or domains of expertise.

This is reflected in the way that city government and related departments classify issues within correspondingly siloed categories such as: "transport", "economic development", "health and public safety", "energy", "housing", "education", etc. These silos or departments typically engage teams of internal or external experts from specific domains of expertise to address a single urban problem (Rockefeller Foundation 2019; Toderian 2015). Architects or urban planners will address public housing or traffic issues, health experts might address issues related to childhood obesity or a rise in type 4 diabetes, while other types of experts tackle issues related to transport or even crime. The solutions that arise out of these "single-solution-to-a-single-problem" approaches rarely shift or resolve a wicked problem because they often mistake *aspects* or *symptoms* for the problem's roots. This top-down, fragmented perspective of a wicked problem is at odds with the bottom-up experience which is always holistic and contextual.

As an example, citizens might experience the urban wicked problem of air pollution most directly in the following ways: increased incidence of respiratory illness, the decline of pollinators (which affects local farmers and gardeners) and loss of green areas within the city (due to arboreal diseases). However, a less direct effect might be an increase in heavy metals and other pollutants in drinking water (via factories that simultaneously pollute air and waterways). Often all of these problems might be addressed individually by different experts and their interconnections, missed.

A traditional governmental solution is to pass policies regulating emissions of all kinds (e.g., factories, cars, coal-fired power stations) and/or to set regional targets for emissions reductions. However, these intended solutions are often connected to or create other problems which may remain invisible to the domain of expertise implementing the solution: unemployment due to companies increasing or passing along costs of goods or services or closure of mines, rising costs of gasoline and energy, all of which disproportionately affects lower-income residents. Furthermore, the adverse health and environmental effects of years of air pollution would not be seen, addressed or remediated by experts working within a single department, domain or silo of expertise because of the multiple boundaries the problem spans (Rockefeller Foundation 2019).

Wicked problems manifest in place, cultural and ecosystem-specific ways

A particular wicked urban problem is *always* connected to other wicked problems, forming complex "clusters" that manifest in ways that are distinct to place, culture and ecosystem. The example of a lack of access to clean water within a city can be connected to a variety of other issues such as: climate change and drought, limited sources of freshwater, a lack of infrastructure to purify water, pollution of sources of freshwater and many more. The way in which these "adjacent issues" inflect the wicked problem is *always* distinct to place.

For instance, a lack of clean water in cities like Ojai California or Phoenix, Arizona is connected to the wicked problems of climate change (drought) and the fact that both cities have limited access to primary sources of fresh water (BBC News 2018). Another factor affecting Phoenix as well as both Bangalore and Beijing is rapidly increasing populations (ibid). However, the problem of water in Bangalore is not only due to rapid growth (and development) but an aging and inadequate infrastructure and sewage system which contributes to water pollution. Beijing also suffers from water scarcity at a higher systems level; China is home to 20% of the world's population, yet has only 7% of the world's fresh water. In addition, 40% of their surface water is too polluted to even use for agriculture or industry (ibid).

Cairo's water shortage is connected to pollution in the Nile from both agricultural and residential waste. Many cities in Africa have a shortage of drinking water connected to the wicked problem of inequality of women and girls (cultural norms), who must often trek for miles to reach water that is unsafe to drink (connected to the wicked problems of a lack of infrastructure and poverty), then carry a limited supply by hand back to the city or village. Jakarta is threatened by rising sea levels which pollute freshwater sources which has led to residents digging illegal wells which are depleting water tables. Water tables in turn are not being replenished because of the wicked problem of overdevelopment has created a prevalence of concrete and asphalt that prevents rainfall from being absorbed (ibid).

Difficulties in framing and addressing wicked problems

The characteristics of wicked problems mentioned earlier make them extremely difficult to see, frame and solve for the following reasons:

1. **Problem Frame:** Because wicked problems are connected to each other at different levels of scale, over multiple time horizons, *and* span disciplinary and professional boundaries, it is difficult to appropriately frame them. Bardwell has noted (1991) that the way in which problems are framed determines how they will be understood and acted upon. Traditional problem-solving approaches favor tight, simplified problem frames which offer the promise of a definitive solution that can be achieved quickly, efficiently and profitably. Framing a complex, wicked systems problem as if it were tame/simple leads to solutions that: fail, unintentionally exacerbate the problem or create new problems elsewhere. Trying to solve a complex systems problem with a non-systemic, cause-and-effect approach is one of the reasons many solutions to wicked urban problems fail.
2. **Problem Context:** Framing complex problem(s) within overly simplified contexts is a related issue. The type of problems discussed in this chapter took an extremely long time to become "wicked", so their evolution within spatio *and* temporal contexts must be considered. The radically large contexts for wicked problems are socio-technical-ecological[6] systems. Socio-technical systems are TANGLES of living and designed/mechanistic systems in which technology plays an ever-increasing role, and these are embedded within the natural world (ecosystems) (Grin, Rotmans & Schot 2010; Trist & Murray 1993; Trist, Emery & Murray 1997). Socio-technical systems are permeated by complex webs of relationship (interactions between people, physical artifacts and infrastructure and the natural world) and these systems, as well as the wicked problems embedded within them, are in a constant state of transition.

    Socio-technical systems tend to become entrenched (locked in) and path-dependent[7] as they evolve *over time*. Understanding how wicked problems arise and evolve within the radically large context of a socio-technical-ecological systems transition is critical in both understanding the problem *and* conceiving solutions to it. It is equally important to consider the trajectory of the systems transition, which includes the near-, mid- and long-term future as context; the context for a wicked problem therefore includes the past (its evolution), it's present (how it manifests at multiple systems levels) and future (its trajectory toward the long-term future).
3. **Unintended Consequences of Solutions:** Both the problem and its larger context can be viewed as a single, large, constantly evolving system within

which the dynamics discussed earlier are at work. When solutions are attempted, the system reacts to these perturbations (system interventions) in unpredictable ways because of its nonlinear dynamics. Most problem-solving approaches are highly linear (simple cause-and-effect principle) and are based upon predictable outcomes that do not take into account nonlinear systems dynamics that are *always* present. The result is that some solutions, when imposed upon the system (perturbations), can lead to unintended and unexpected negative consequences. The ramifications of *past* interventions may manifest in the *present* in ways that exacerbate the problem or create new ones. For these reasons, it is often difficult to distinguish between the genuine "root causes" of a wicked problem and its "consequences". It is therefore necessary to understand the manifestation of the problem in the present, but also its historical origins, and evolution.

4. **Solutions versus Systems Interventions:** When both the wicked problem *and* its spatio- temporal-ecological context are seen as a large system that is constantly transitioning and evolving, it becomes clear that a single, one-off solution cannot resolve it. Instead, the paradigm must shift from "solving a problem" (one-off solutions), to "intervening in a system". Interventions are made at multiple levels of scale (household, neighborhood, city and region) (Kossoff 2011, 2019a) over multiple time horizons. Some interventions are short-lived, while others might continue for years or decades. Interventions (multiple solutions aimed at wicked problems) are intended as perturbations that destabilize entrenched systems, and begin to "nudge" the socio-technical system's transition trajectory toward a new, more desirable long-term future. In this way, interventions go beyond the resolution of a single wicked problem or problem "cluster" to that of designing *for*, or catalyzing an entire "system transition". A useful change in metaphor for the problem-solving process is to shift from "solving problems" to "solutioning over time" and realizing that aiming for a final, clear resolution to a problem is unrealistic. Instead, intervening becomes an ongoing "solutioning" process that resembles the design of software. Each new version is a solution that moves the application forward by solving for problems or implementing new capabilities. It is a process, not a destination.

5. **Systems Problems Take a Long Time to Resolve**: Because the problems that arose in socio-technical-systems transitions took a long time to become wicked, it will take a long time to resolve them and shift the transition trajectory. Traditional problem-solving approaches are predicated upon clear problem definitions that lead to rapid and cost- effective solutions. The focus on siloed solutions for individual problems within short time spans is at odds with wicked problem resolution. Because government agencies and funding institutions still adhere to these approaches, few projects or initiatives are funded long enough for wicked problems to be resolved and for systems to shift their transition trajectories. Funding paradigms will need to transform with the objective of funding problem resolution over multiple years or decades (myriad systems interventions at different levels of scale, over

multiple time horizons). An opposite approach is merited; funding multiple interventions aimed at resolving an ecology of wicked problems over multiple years or decades. This will not only resolve multiple complex problems more quickly, but will also destabilize entrenched socio-technical systems and enable transitions toward more sustainable, long-term futures.

The need for a new systems approach for solving wicked problems

Transition design is a new, transdisciplinary approach for solving complex, wicked problems that focuses on the need for entire societies, cities and communities to transition toward more sustainable and equitable long-term futures (Irwin 2011a, 2011b; Kossoff 2011; Irwin 2015; Irwin, Kossoff & Willis 2015; Irwin, Tonkinwise & Kossoff 2015; Kossoff, Tonkinwise & Irwin 2015; Irwin, Tonkinwise, Kossoff & Scuppelli, 2015; Tonkinwise 2015a, 2015b; Mulder & Loorbach 2016; Boehnert 2018; Escobar 2018; Auger 2019; Irwin 2019; Irwin & DiBella 2019; Boylston 2019; Gaziulusoy 2019; Hanington 2019; Kossoff 2019a, 2019b; Tonkinwise 2019; Ceschin & Gaziulusoy 2020; Irwin & Kossoff 2020). Transition design was developed in response to the urban challenges discussed earlier and aspires to transcend traditional one-solution-for-a-single-problem approaches that are inadequate in dealing with complex, systemic problems. Transition design emphasizes:

- The need to frame problems within radically large, spatio-temporal contexts that include the past (how the problem evolved over long periods of time), present (how the problem manifests at different levels of scale) and future (visions of the long-term future in which the problem has been resolved).
- The need for the stakeholders affected by the problem(s) to be involved throughout the problem framing, visioning and solutioning process. This challenges many dominant processes in which professional or disciplinary experts from the outside problem solve/design "for" the communities affected by the problem(s). Transition design aspires to continually leverage the knowledge and wisdom from inside the system and build community capacity to self-organize, advocate and problem solve (Carlsson-Kanyama *et al.* 2008; Baur *et al.* 2010; Simon & Rychard 2005; Dahle 2019).
- The need for stakeholders to co-create long-term visions of desirable futures, as a way to transcend their differences in the present and focus on a future space in which they are more likely to agree.
- The need to create "ecologies of synergistic interventions"[8] (solutions) that are connected to each other and the long-term vision as a strategy for transitioning entire societies toward a desirable, long-term futures.
- The need to think and work for long horizons of time.[9] Resolving wicked problems and transitioning entire societies toward sustainable long-term futures will unfold over many years or even decades and will require patience,

tenacity and an ongoing process of visioning and solutioning to remain on course during the transition.

The transition design framework

The transition design framework (Figure 5.1) (Irwin 2015) comprises four mutually influencing and co-evolving areas of practices, knowledge and skill sets relevant to understanding, seeding and catalyzing systems-level change.

> **Mindset and Posture:** Living *in* and *through* transitional times calls for self-reflection and new, more holistic ways of "knowing" and "being" in the world. Fundamental change is often the result of a shift in mindset or worldview that in turn leads to new modes of behavior, action *and* interaction

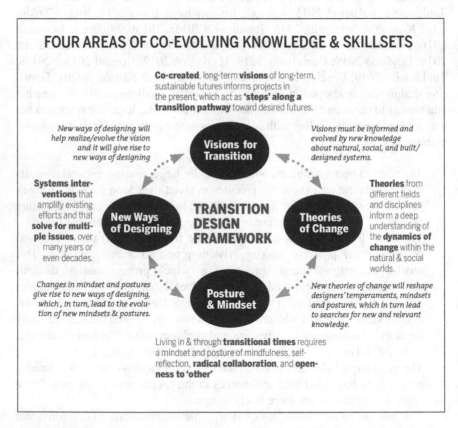

FOUR AREAS OF CO-EVOLVING KNOWLEDGE & SKILLSETS

Co-created, long-term **visions** of long-term, sustainable futures informs projects in the present, which act as **'steps' along a transition pathway** toward desired futures.

New ways of designing will help realize/evolve the vision and it will give rise to new ways of designing

Visions for Transition

Visions must be informed and evolved by new knowledge about natural, social, and built/designed systems.

Systems interventions that amplify existing efforts and that **solve for multiple issues**, over many years or even decades.

New Ways of Designing

TRANSITION DESIGN FRAMEWORK

Theories of Change

Theories from different fields and disciplines inform a deep understanding of the **dynamics of change** within the natural & social worlds.

Changes in mindset and postures give rise to new ways of designing, which, in turn, lead to the evolution of new mindsets & postures.

Posture & Mindset

New theories of change will reshape designers' temperaments, mindsets and postures, which in turn lead to searches for new and relevant knowledge.

Living in & through **transitional times** requires a mindset and posture of mindfulness, self-reflection, **radical collaboration**, and **openness to 'other'**

Figure 5.1 The transition design framework comprises four mutually influencing and co-evolving areas of knowledge and skill sets relevant to understanding how change happens in complex systems.

with others (Meadows 1999). Individual and collective mindsets represent the beliefs, values, assumptions and expectations formed by our individual experiences, cultural norms, religious and spiritual beliefs and the socio-economic and political paradigms to which we subscribe and are situated within (Crompton 2016). These mindsets and postures often go unnoticed and unacknowledged but they profoundly influence *if* and *how* we see a problem, as well as how we frame (setting context) and solve it (Irwin 2011a). Transition design examines the phenomena of worldview, posture and mindset and their connection to wicked problems and proposes that shifting values and postures can be part of an intentional process of self-reflection and change (Boehnert 2019).

Theories of Change: Transition design argues that the social, economic, political and technological systems upon which society depends *must* transition toward more sustainable long-term futures. Seeding and catalyzing change within complex systems will require a deep understanding of the nature of change itself – how it manifests and how it can be intentionally directed (Eguren 2011). Understanding theories of change is essential because: 1) a theory of change is *always* present within a planned/designed course of action, whether it is explicitly acknowledged or not; 2) transitions toward sustainable futures will require sweeping change at every level of our society; 3) conventional, outmoded or incorrect ideas about change lie at the root of many wicked problems.

Vision: more compelling, co-created (by stakeholders affected by the problem(s)) visions of long-term sustainable and equitable futures are needed. Drawing upon a variety of foresighting and visioning approaches, stakeholders affected by the problem(s) co-create compelling visions of long-term, place-based lifestyles in which the problem has been resolved (preferred futures) (Candy & Kornet 2019; Candy & Lockton 2019). These visions enable stakeholders to transcend their differences in the present and enter into a creative space in which they explore possibilities that they can agree on. These co-created and continually evolving visions act as both "magnets", motivating action in the present, as well as a "compass" or roadmap for how to transition toward the desired future (Irwin 2019).

New Ways of Designing: The transition to a sustainable society will require new problem-solving approaches informed by different value sets and knowledge. Addressing systems problems will require "ecologies" of systems interventions (solutions) implemented at multiple levels of scale, along multiple time horizons. An intentionally short-lived solution might act as a step toward a longer-term goal, while others might be designed to change and evolve over long periods of time. A key strategy is to connect ecologies of solutions to each other *and* the long-term vision(s) for greater traction and leverage (destabilizing "stuck" systems and shifting their trajectory toward the desired future). Transition designers also look to amplify efforts, and solutions already underway and integrate them as part of the ecology of solutions (Irwin 2019).

The transition design approach and its objectives

Transition design should be considered an "approach" (see Figure 5.2) for addressing wicked problems and seeding and catalyzing societal transitions toward more sustainable futures, as opposed to a templatized, linear process. This is because wicked problems *must* be addressed in place-based ways that acknowledge the uniqueness of local cultures, socio-economic-political conditions and local eco-systems. The approach sets several objectives that have to do with understanding the problem, framing it within appropriate contexts and taking a systems approach to developing solutions. New tools and methodologies are being developed, but more will be needed. This chapter reports on approaches being developed at the School of Design at Carnegie Mellon University (Rohrbach & Steenson 2019; Irwin & Kossoff 2020).

Understanding the problem and who it affects: mapping wicked problems and stakeholder relations

Two mapping steps are interconnected: 1) mapping the problem and 2) mapping stakeholder relations. Each informs the other; therefore a sequential process is not suggested. Ideally, a problem map should be highly visual and built up *over time* by the stakeholders themselves.[10] An initial problem map created by stakeholders can serve as a "sketch" for further research (using a variety of research methods to reach *all* groups affected by the problem) to validate or refute the perspectives captured in the initial mapping stage.

At the outset of problem mapping, it may be difficult to identify all stakeholder groups affected by a problem(s) and additional groups are often revealed throughout the mapping process. This is why a "back and forth" between problem and stakeholder mapping is useful. Both maps serve as visual representations/reservoirs of an accruing body of knowledge about the problem(s) and the stakeholder groups affected by it.

MAPPING THE PROBLEM leverages the diversity of perspectives, knowledge and expertise related to the problem that resides *within the system*. The objective is that either via workshops or field research, all stakeholders identify as many issues related to the problem(s) as possible, within 5 key areas: 1) social issues (how/what stakeholders think and do); 2) infrastructure and technology issues; 3) economic and business issues; 4) policy, governance and legal issues; 5) environmental issues. The process of building a comprehensive problem map may take months of direct interaction with stakeholders, scaffolded by both desktop and field research to ensure accuracy. The map is conceived as a continually

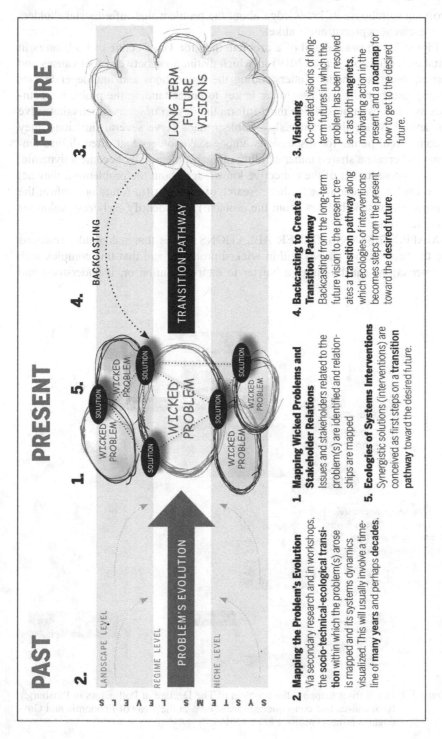

Figure 5.2 Overview of an emerging transition design approach.

evolving storehouse of knowledge about the problem that informs stakeholders and researchers/practitioners alike.

Figure 5.3 shows a detail of a problem map for The Decline of Pollinators in Pittsburgh (Guilfoile *et al*. 2019) in which distinctions between root causes and consequences are made. Understanding the connections and interdependencies among issues in the five categories is key to understanding the problem's complex systems dynamics which must inform the design of interventions that solve for several issues simultaneously. Problem maps serve several functions: they become focal points for discussions among stakeholders that often diffuse tensions and create a shared understanding of the problem; they become a dynamic, visual representation of the collective knowledge about the problem(s); they act as an evolving agenda for further research of all kinds (to refute or confirm the perspective of stakeholders about the problem); can identify early/easy solutions (interventions).

MAPPING STAKEHOLDER RELATIONS argues that stakeholder relations are the "connective tissue" within wicked problems and that this complex web of interactions can be either a barrier to their resolution or, if understood and

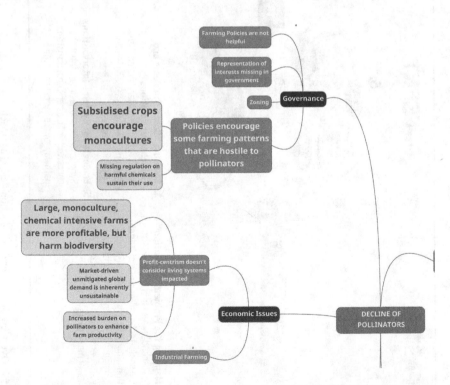

Figure 5.3 Detail from mapping the problem of The Decline of Pollinators in Pittsburgh. Root causes and consequences are shown in the areas of Economic and Governance issues, Guilfoile *et al*. (2019).

leveraged, a boon to problem resolution. The objective is to identify *all* of the stakeholder groups affected by the problem; both human and nonhuman, living and nonliving, then try to understand the nature of the relations among different groups.

Several approaches have been tested (Irwin & Kossoff 2017a, 2017b; Hamilton 2019) in workshop settings. The Stakeholder Triad exercise asks stakeholders themselves to formulate a comprehensive list of all of the groups affected by the problem, then identify the three groups most likely to disagree. Next, stakeholders speculate on what the areas of conflict between the three groups might be (relative to the problem) as well as areas in which they may be in alignment. A discussion follows, focused on the ways in which areas of conflict can be *barriers* to problem resolution, but conversely, how areas of agreement can be leveraged toward *resolution*. Another exercise conducted in workshops asks each stakeholder group

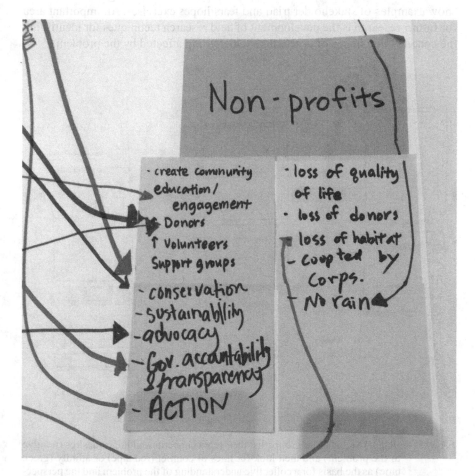

Figure 5.4 (Left) Detail of a stakeholder triad mapping exercise in which lines of affinity or conflict between groups are mapped.

to list their fears and concerns relative to the problem (on pink paper), as well as their hopes and desires (on green paper). Each group places their lists on a wall and are then invited to use red and green tape to connect lines of conflict (red) and agreement (green). Groups are asked to particularly note instances in which one group's greatest fear could be another group's fondest hope.

This informal and rather "boisterous" process interjects an element of "play" into a potentially tense and confrontational debate among diverse stakeholder groups about how to solve the problem. Like the problem mapping exercise, it does not produce reliable, qualitative data or concrete conclusions, but it does serve several important functions: 1) begins a dialog among diverse stakeholder groups who do not agree; 2) enables them to experience the problem's complexity firsthand and can produce the realization that there is no single solution to the problem and 3) can potentially serve as a sketch for further structured research to validate or refute outcomes from the workshop (Irwin 2019). Figures 5.4 and 5.5 show examples of stakeholder triad and fears/hopes exercises. An important area for further research is the development of field research techniques for identifying the concerns and hopes of *every* stakeholder group affected by the problem.

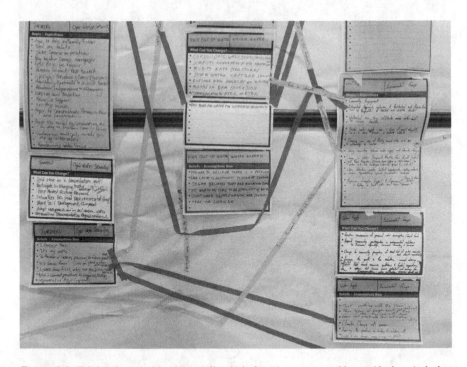

Figure 5.5 (Right) Stakeholder groups list their fears/concerns and hopes/desires (relative to the problem) and then look for lines of conflict (red tape) or affinity (green tape) as the basis for a collective understanding of the problem and the perspectives which contribute to its intractability.

Understanding how the problem evolved (over a long periods of time)

Wicked urban problems arise *over time* within complex, socio-technical-ecological systems as the result of highly complex interactions at different levels of scale between people, the built world (artifacts, infrastructure, laws/guidelines, scripted interactions, etc.) and the natural environment. Transition design argues that an understanding of the historical evolution of the problem within a large, spatio-temporal frame can inform more appropriate and synergistic interventions (solutions) in the present.

Transition design draws upon the Multi-Level-Perspective framework (MLP) from socio-technical transition theory (Grin *et al*. 2010), which explains how change happens within socio-technical systems at different, interconnected levels. The MLP identifies three distinct systems levels in which events unfold, infrastructure and artifacts arise, and webs of interaction occur: the macro or **Landscape** level, where large, slow moving change occurs; the meso or **Regime** level where societal infrastructure and established ways of doing things occur; and the micro or **Niche** level where innovations or disruptions to the status quo are incubated. Systems interactions are social, technical, institutional, infrastructural and normative and the *networks of relationship* within and between systems levels (and their various actors/factors) become progressively more "entrenched" and resistant to change as their scale and duration over time increases. Eventually, large systems (and the wicked problems within them) become "locked in" to a particular trajectory or transition pathway (societal transitions toward increasingly unsustainable, long-term futures). Figure 5.6 shows an example of an MLP problem evolution map for *A Lack of Access to Healthy Food in Pittsburgh*.

Tools for mapping the historical emergence and evolution of wicked problems for use in workshops and as a guide for conducting secondary research are being developed and more are needed. Understanding how a problem emerges in the long-term past and understanding the complex systems dynamics that exacerbated it (sometimes over the course of decades) can lead to insights that are important in understanding how it manifests in the present and even more importantly, why it may have become intractable. Of equal importance is to understand *what was happening in the socio-technical system before the problem arose*. What conditions were in place or what was absent that prevented the problem from arising? How were people living? What cultural norms, beliefs, behaviors and practices kept the problem from arising?

Students in the Carnegie Mellon Transition Design seminar conducted secondary research on the problem of *The Rise of Air Pollution in Pittsburgh* and found that in the present and recent past, there was a relatively high tolerance of the problem. Just one facet of the historic socio-technical transition that gave rise to the problem showed that Pittsburgh's long-time status as a center for steel production and coal mining (both of which created some of the worst air pollution in the nation's history) (Karavdic *et al*. 2019) led to collective attitudes that darkened skies from air pollution was equated with a booming economy and prosperity.

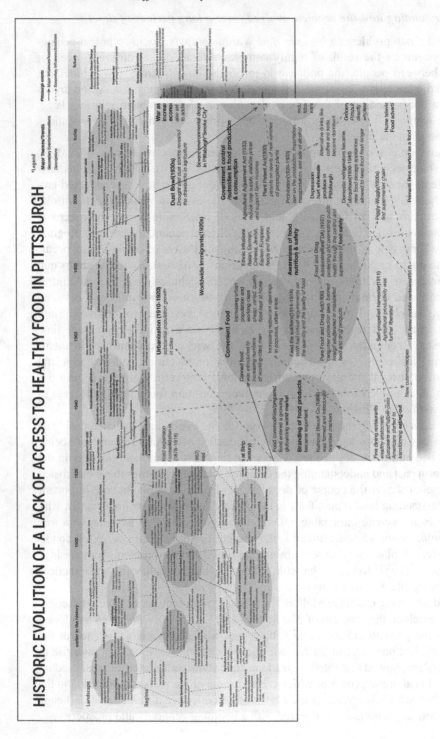

Figure 5.6 An MLP problem map showing the evolution of the problem of A Lack of Access to Healthy Food in Pittsburgh. The map covers a time span of several decades and the chronicles changes at three levels within the socio-technical system transition within which the problem evolved (Singh *et al.* 2019).

This collective belief/social norm at the Landscape level, contributed to the intractability at the Regime level through a lack of public and political will to pass legislation regulating pollution levels and remediating adverse health effects. Figure 5.7 shows a detail of the problem evolution map showing events and dynamics between the three system levels.

The second objective in mapping the historic evolution of problems within transitioning socio-technical-ecological systems is to enable teams and community members to acquire a "feeling for systems dynamics" that must inform interventions in the present and near-term. Such interventions can simultaneously solve

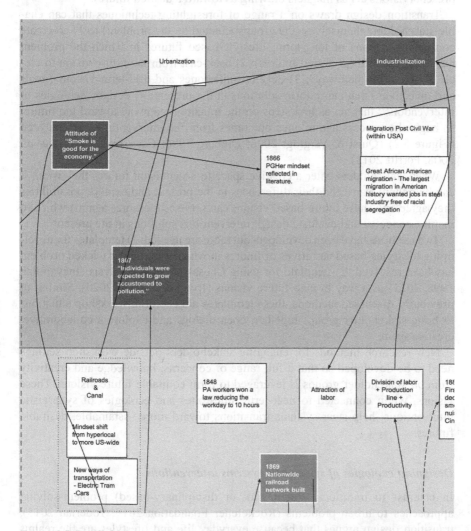

Figure 5.7 Detail from a problem evolution map for Poor Air Quality in Pittsburgh that shows how collective attitudes about air pollution at the landscape level led to a higher degree of tolerance for air pollution (Karavdic *et al.* 2019).

for wicked problems *and* destabilize entrenched systems; intentionally "nudging" their trajectories toward more sustainable, long-term futures.

Visioning and back-casting

A primary objective of the transition design approach is to seed and catalyze societal transitions toward more equitable and sustainable long-term futures. In order to achieve this, more compelling visions of sustainable futures must be developed to guide these transitions (visions act as roadmaps) and motivate action in the present (visions act as magnets drawing us toward a desired future).

Transition design draws on a range of foresighting techniques that can enable stakeholders themselves (the groups affected by the problem) to 1) co-create compelling visions of long-term, lifestyle-based futures in which the problem has been resolved (preferred futures); 2) back-cast from the future vision to create a "transition pathway"; 3) develop milestones and mid-term visions (goals and objectives) that can inform solutions in the present; 4) design "ecologies of interventions" that act as first steps on the transition pathway toward the future; 5) begin the process again, once outcomes from the interventions are analyzed (Figure 5.2) (Quist & Vergragt 2006; Dreborg 1996; Carlsson-Kanyama *et al.* 2008; Porritt 2013).

When stakeholders enter a creative space to co-envision futures they *want*, it can help them transcend their differences in the present and explore possibilities they agree on. These future-based visions can serve as measures against which to inspire, conceive and evaluate design interventions/solutions in the present.

Two methods have been developed, but more are needed: 1) templates for developing lifestyles-based narratives of futures in which a particular wicked problem has been resolved; 2) template for using Causal Layered Analysis (Inayatullah 1998, 2013) as a way to map future visions (Irwin & Kossoff 2020). Similar to previously discussed methods, these templates are useful in workshop situations to bring stakeholder groups together, open dialogs and explore a collaborative, creative space.

New research methods for engaging stakeholders outside workshop settings need to be developed so that a full range of concerns, knowledge and creativity from *all* stakeholder groups is leveraged to form equitable future visions. These visions, when connected to near-term milestones and ecologies of synergistic interventions, help steer systems transitions toward more sustainable, desirable futures.

Designing ecologies of synergistic systems interventions

In contrast to traditional (often silo- or disciplinary-based) problem-solving approaches to urban problems (Rockefeller Foundation 2019; Toderian 2015), transition design argues that because everyday life and lifestyles are the realms within which the consequences of wicked problems are *experienced*, it should be the primary context within which to address them (Kossoff 2011; Debord 2002).

Transition design emphasizes the co-creation of lifestyle-based visions of the long-term future that can inform solutions/interventions in the present. Transition design systems interventions are characterized by:

- Everyday life and lifestyles are seen as the most important and fundamental context for design.
- Solve for short, medium and long horizons of time, at all levels of scale of everyday life (the household, the neighborhood, the city, the region) (Kossoff 2011, 2019a).
- Look for emergent possibilities within problem contexts and amplify grass-roots efforts and solutions that are already underway.
- Are part of an "ecology of solutions" that are linked to each other *and* the long-term vision to act as steps on a "transition pathway" toward the desired future.
- Being both material/tangible as well as immaterial/intangible. Interventions acknowledge the ways in which worldviews, mindsets, beliefs, assumptions, values, behaviors and practices can be used as leverage points for change (Meadows 1999) in complex systems.
- Distinguishing between "wants" or "desires" and genuine needs and base solutions upon maximizing the *satisfiers* for the widest possible range of needs.

Key knowledge sets and theories that inform transition design solutions include: Manfred Max-Neef's theory of needs and satisfiers (Max-Neef 1991); practice theory (Shove & Walker 2010; Shove *et al.* 2012; Kuijer & De Jong 2011); design for behavior change (Lockton 2014; Niedderer, Clune & Ludden 2017); values and frames (Crompton 2016); design for the circular economy (Chapman 2017; Ellen MacArthur Foundation 2017; Royal Society of Arts 2013); commoning and the sharing economy (Tonkinwise 2018; Bauwens 2017; Gruber 2018; Thackara 2015; Harvey 2012); Causal Layered Analysis (Inayatullah 2013, 1998); and the Domains of Everyday Life Framework (Kossoff 2011, 2019a).

Looking at a "cluster" of interconnected wicked problems

The transition design approach argues that wicked urban problems do not arise and evolve in isolation from one another; rather they couple with other wicked problems to form "problem clusters". These tightly coupled clusters contribute to each problem's intractability and collectively are barriers to systems change and transition. However, the areas of overlap and interconnection *among* problems become "zones of possibility" in which "ecologies of synergistic interventions" have the potential to address multiple problems and issues simultaneously (Figure 5.2, step 5).

As an example, Figure 5.8 shows how the problem of Homelessness in Pittsburgh manifests in ways that are unique to the city and connect to a host of other urban problems. Although homelessness is a problem shared by most urban areas

Figure 5.8 Wicked urban problems always manifest in place and culture-based ways and are usually connected to other wicked problems to form problem "clusters". Here the problem of homelessness in Pittsburgh, PA, USA, is connected directly to the high cost of housing, poverty and opioid addiction.

around the world, the way in which it manifests is *always* distinct to a particular place, its culture and its historic evolution (often over many years or decades).

In Pittsburgh, Pennsylvania, the problem of homelessness is tightly coupled to the problems of opioid addiction (the state has one of the highest addiction rates in the country) (CDCP n.d.; ARC and NaCo 2019), the increasing cost of housing and poverty in general. Extreme poverty in Pittsburgh is a legacy of the demise of manufacturing in rust belt[11] cities in the United States in general, and the demise of the steel mill economy in Pittsburgh in particular (Haller 2005). Similarly, one of the roots of opioid addiction is the decades-long practice of overprescribing for a blue-collar workforce that was/is highly susceptible to injuries from large machinery and repetitive stress (ARC and NACo 2019).

The demise of Pittsburgh's steel industry resulted in its population decreasing by almost 1/2 (50%) between the years of 1950 and 1990 (Haller 2005), which exacerbated the problems of unemployment and homelessness. The response from city leaders has been a multi-decade effort to reinvent and rejuvenate the city (Russo 2017) which has led to urban growth, the attraction of high-tech industry to Pittsburgh, the inevitable rise in cost of residential housing, gentrification and the commoditization of housing. These factors, in combination with Pittsburgh's history of racism, create a problem cluster that is unique.

As previously discussed, most city or regional government departments and non-profits take a siloed approach to addressing wicked urban problems like homelessness, and a review of solutions implemented in Pittsburgh and similar cities fall into distinct categories:

The Band-Aid or Temporary Fix: These types of solutions usually address a *facet* of the problem or a particular situation. This might include a fund drive for clothing for the homeless or temporary shelters that have limited capacity. They help a limited number of individuals for a limited period of time, but do not create positive, systemic change.

Rehabilitative Solutions: With respect to homelessness, this type of solution might involve helping homeless people get off the streets by providing guaranteed transitional housing for a limited period of time which enables them to recover/rehabilitate and perhaps find permanent housing and employment. These solutions have limited success.

Preventative Solutions: These solutions are aimed at preventing homelessness in the first place and focus on what causes people to become homeless. Solutions might include legal help for people at risk of eviction from their homes, an affordable rental housing locator or rent allowance schemes.

Systemic Solutions (interventions): These solutions are aimed at ongoing, systems-level change and often happen at the level of policy and governance as solutions that aim to shift perceptions, mindsets and values. For example, a policy that requires new housing to be mixed income, "housing first" initiatives that argue that housing is fundamental and issues such as substance abuse must be dealt with after a homeless person has shelter (Pohjanpalo 2019), universal allowance for all citizens (Haagh & Rohregger 2019), campaigns to shift public attitudes about homelessness and projects to provide transport to school for children of homeless families so that their education is not interrupted. In contrast to other types of solutions, these projects and initiatives have the potential to spark systems-level change when connected to each other and long-term visions, *consistently over time*.

Ecologies of Synergistic Solutions (interventions): These types of solutions are rare but have the greatest potential to seed and catalyze systems level change because they are aimed at the resolution of several problems at once. These are solutions/interventions situated within "zones of opportunity" within a problem cluster (areas where they overlap). Instead of asking what a solution to homeless might be, transition designers ask "what are interventions that solve simultaneously for homelessness, opioid addiction, the high cost of housing and poverty?" Questions like these are rarely asked because of the siloed approaches discussed earlier and the difficulty politicians and other experts have in collaborating across these siloes. Another key aspect of the strategy is to conceive solutions/interventions that are situated at different scales of everyday life (Kossoff 2011): the household, the neighborhood, the city and the region, to ensure that interventions are happening at all systems levels.

In the example of homelessness, creating an ecology of synergistic solutions/interventions might involve connecting *existing* projects in different problem sectors to each other at different levels of scale, then asking where gaps are within the zones of opportunity. These gaps are often ideal places for the incubation of new interventions that can join an ecology to become a step on a transition pathway toward a co-created, desirable future. It is important to emphasize that solutions/ interventions will be both material (artifacts, technology, infrastructure and scripted interactions) as well as nonmaterial (shifting attitudes, beliefs, values, perceptions, behavior and practices) (Lockton 2014; Crompton 2016; Shove & Walker 2010; Shove *et al.* 2012).

Creating a hypothetical ecology of interventions, as shown in Figure 5.9, will require new transdisciplinary/cross-sector approaches: stakeholders, representatives from multiple government departments, cross-sector experts and funding organizations will need to come together to conceive new solutions that connect

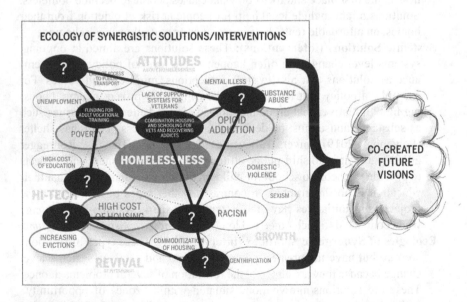

Figure 5.9 Existing and new projects and solutions within the entire wicked problem cluster are connected to each other and the co-created, long-term vision, to form an ecology of synergistic solutions/interventions that address multiple problems and issues simultaneously.

to each other, existing projects and initiatives *and* co-created, long-term visions. Together they will form ecologies of synergistic interventions aimed at resolving entire wicked problem clusters and catalyzing the transition of the socio-technical-ecological systems which forms their larger context.

Conclusion

The strength that comes from human collaboration is the central truth behind civilization's success and the primary reason why cities exist.
                                                                      – Edward Glaeser (2011)

This chapter has argued that many of the complex wicked problems confronting 21st century societies manifest at the level of the city in ways that negatively affect millions of people. It also argues that these problems are systems problems that are interconnected, resistant to resolution (intractable) and collectively are directing the transition of entire societies (socio-technical-ecological systems) toward unsustainable, long-term futures. But transition design argues that cities *themselves* can become powerful "leverage points" for systems-level change and societal transitions. Transition design is proposed as a strategy that can address multiple wicked problems simultaneously (destabilizing entrenched socio-technical systems) and intentionally shift transition trajectories for entire communities toward more sustainable, equitable and desirable futures.

Transition design's aim is to develop new narratives, tools, strategies and methodologies that support the current trend towards decentralization and re-localization that is occurring at the level of the city, and is often led by municipal authorities (Barber 2013; Glaeser 2011; Katz & Nowak 2017). This trend could inform an intentional transition toward local, regional and planetary networks of cities working together, experimenting with multi-local, place-based solutions, to address their common problems (Kossoff 2019a; Kostakis *et al.* 2015; Manzini 2011). This trend is of particular importance at a time when many nation-states are taking up isolationist stances, have been compromised by special interests and are failing to take responsibility for local affairs (Barber 2013).

The transition design approach could be embedded in city halls, community centers, local councils, businesses, foundations, NGOs, charities, hospitals, universities and schools. But, it will require new structures and processes – a reinvigoration of local democracy and new kinds of digital platforms – through which all stakeholders can ongoingly participate in addressing the wicked problems that they are affected by.

Notes

1 By "design-led", we do not mean that professional designers will lead transition/
   problem-solving efforts, rather that transdisciplinary teams and affected communities
   will use many of the tools and approaches *used* by professional designers, such as
   iterative prototyping, qualitative/user-centered research approaches and materializing
   solutions.

2 Emergence refers to the unpredictable and new forms of physical and behavioral order within complex systems (and wicked problems). These forms arise at critical points of instability within the system that are a response to perturbations from the environment. Therefore, complex systems cannot be understood through reductive analysis in which the sum of the individual parts explains the behavior of the whole; the whole cannot be understood independent of the context within which it exists. The relevance for transition designers in relation to emergence, self-organization and sensitivity to initial conditions is that the outcomes of systems interventions cannot be predicted and we must wait to see how the system responds before intervening again.

3 Self-organization describes the dynamics at work within complex open systems: they are structurally closed/organizationally open; are in a continual exchange of energy and matter with their environment; they are "self-making" (autopoetic). Their responses to perturbations from their external environment are unpredictable and self-determined; they couple with their environment and co-evolve in symbiotic relationship with it.

4 Sensitivity to initial conditions refers to the way in which living systems display extreme sensitivity to initial conditions; small changes (interventions) have the potential to ramify throughout the system and create large variations in its long-term behavior. Examples are global warming, weather patterns, a viral epidemic. This principle from chaos and complexity science posits that the deterministic nature of these systems (indicated by their initial conditions) cannot be predicted due to the chaotic behavior of the system over time as it is perturbed by small and large changes from its environment. Long-term predictions of its behavior are impossible. A commonly used metaphor is that a butterfly flapping its wings in the Amazon can give rise to a hurricane in southeast Asia.

5 Greywater or sullage is all the wastewater generated in households or office buildings from streams without fecal contamination, i.e., all streams except for the wastewater from toilets. Sources of greywater include showers, baths, washing machines or dishwashers.

6 Socio-technical sustainability transitions theory (STRN) refers to the socio-technical system; however, transition design argues after the Tavistock Institute (Trist, Emery and Murray 1997) which first proposed the term and referred to it as socio-technical-ecological systems in acknowledgement that the built/technical world is embedded with the natural world.

7 The concepts of path dependency and "lock in": the concept of *lock-in* refers to a socio-technical system becoming self-reinforcing in part because of its adoption of a certain technology or "way of doing things" that give incumbent actors and technologies an advantage over new entrants. Path dependency limits the options of the actors, institutions and networks and is a primary cause of intractability and the transition of large socio-technical systems toward unsustainable, long-term futures.

8 Synergistic interventions or solutions are those which address more than one issue or a problem simultaneously. Transition Design argues that within problem "clusters" there are "zones of opportunity" in which synergistic interventions/solutions have the potential to solve for multiple issues simultaneously. It draws from economist Manfred Max-Neef's (1991) theory of "synergistic satisfiers", in which several needs can be satisfied simultaneously.

9 The term "long horizon of time" refers to a timeframe of multiple years, decades or even hundreds of years into the future. It draws on Stewart Brand's argument that one of the roots of complex problems is our societies' propensity to think in ever shorter time horizons (Brand 1999).

10 The problem map or map of a problem cluster has the potential to become an evolving representation of a community's understanding of the problem(s). A city's Transition Design Office might take the form of a main street store in front of which members of the community are invited to come in and interact with the map, adding their knowledge to it and gaining a deeper understanding of the collective perspectives represented in it.

11 Rust Belt is an informal term for a region of the United States that experienced industrial decline starting around 1980. Generally it includes Appalachia (a cultural region
in the Eastern United States that stretches from the southern part of New York State to
northern Alabama and Georgia), the Midwest and Great Lakes.

References

Andersen, S.E. & Nielsen, A.E., 2009, 'The city at stake: Stakeholder mapping the city',
Culture Unbound 1(2), 305–329, viewed 6 February 2020, from www.cultureunbound.
ep.liu.se/v1/a19/cu09v1a19.pdf.

Appalachian Regional Commission (ARC) and National Association of Counties (NACo),
2019, *Opioids in Appalachia: The role of counties in reversing a regional epidemic*,
ARC and NACo, Washington, DC, viewed 7 February 2020, from www.naco.org/sites/
default/files/documents/Opioids-Full.pdf.

Auger, J., 2019, *Cameron tonkinwise: Creating visions of futures must involve thinking
through complexities*, SpeculativeEdu., viewed 1 February 2020, from https://specula-
tiveedu.eu/interview-cameron-tonkinwise/.

Australian Public Service Commission, 2007, *Tackling Wicked Problems: A Public Policy
Perspective*, Commonwealth of Australia, Canberra, viewed 6 February 2020, from
www.enablingchange.com.au/wickedproblems.pdf.

Barber, B., 2013, *If Mayors Ruled the World: Dysfunctional Nations, Rising Cities*, Yale
University Press, New Haven. Rockefeller Foundation (2019), pp. 3–9.

Bardwell, L., 1991, 'Problem-Framing: A Perspective on Environmental Problem-Solving', *In Environmental Management* 15, 603–612.

Baur, V., Elteren, A., Nierse, C. & Abma, T., 2010, 'Dealing with distrust and power
dynamics: Asymmetric relations among stakeholders in responsive evaluation', *Evaluation* 16, 233–248.

Bauwens, M.*et al.*, 2017, *A Commons Transition Primer*, The Peer2Peer Foundation,
Amsterdam, viewed 2 February 2020, from https://primer.commonstransition.org.

BBC News, 2018, *The 11 cities most likely to run out of drinking water – Like Cape Town*,
viewed 7 February 2020, fromwww.bbc.com/news/world-42982959.

Boehnert, J., 2018, *Design, Ecology and Politics. Towards the Ecocene*, pp. 22–26,
Bloomsbury, London.

Boehnert, J., 2019, 'Transition design and ecological thought', in T. Irwin & D. DiBella (eds.),
Cuaderno73:Designinperspective:Transitiondesignmonograph,pp.133–148,University
of Palermo, Buenos Aires, viewed 7 February 2020, from www.academia.edu/36760846/
Cuaderno_Journal_73_Design_in_Perspective_Transition_Design_Monograph.

Boylston, S., 2019, *Designing with society: A capabilities approach to design, systems
thinking and social innovation*, Routledge, Abingdon, pp. 12, 99, 146.

Brand, S., 1999, *The Clock of the Long Now: Time and Responsibility*, Basic Books, New
York, NY.

Briggs, J. & Peat, D., 1989, *Turbulent Mirror: An Illustrated Guide to Chaos and the Science of Wholeness*, Harper and Row, New York, NY.

Buchanan, R., 1992. 'Wicked problems in design thinking', *Design Issues* 8(2), 5–21.

Byrne, D., 1998, *Complexity Theory and the Social Sciences: An Introduction*, Routledge,
Abingdon.

Candy, S. & Kornet, K., 2019, 'Turning Foresight Inside Out: An Introduction to Ethnographic Experiential Futures', in S. Candy & C. Potter (ed.), *Design and Futures*,
pp. 3–21, Tamkang University Press, Taipei.

Candy, S. & Lockton, D., 2019, 'A vocabulary for designing for transitions', in T. Irwin & D. DiBella (eds.), *Cuaderno 73: Design in perspective: Transition design monograph*, pp. 27–49, University of Palermo, Buenos Aires, viewed 7 February 2020, from www.academia.edu/36760846/Cuaderno_Journal_73_Design_in_Perspective_ Transition_Design_Monograph.

Capra, F. & Luisi, P.L., 2016, *The Systems View of Life: A Unifying Vision*, Cambridge University Press, Cambridge.

Carlsson-Kanyama, A. *et al.*, 2008, 'Participative Backcasting: A Tool for Involving Stake-holders in Local Sustainability Planning', *Futures* 40, 34–46.

Center for Disease Control and Prevention (CDCP), n.d., *Opioid overdose* [online], viewed 6 February 2020, fromwww.cdc.gov/drugoverdose/data/statedeaths.html.

Ceschin, F. & Gaziulusoy, I., 2020, *Design for Sustainability: A Multi-Level Framework from Products to Socio-Technical Systems*, pp. 124–136, Routledge, Abingdon.

Chapman, J., 2017, 'Product Moments, Material Eternities', in D. Baker-Brown (ed.), *The Re-Use Atlas: A Designer's Guide Towards the Circular Economy*, pp. 161–165, RIBA, London.

Coyne, R., 2005, 'Wicked problems revisited', *Design Studies* 26, 5–17.

Crompton, T., 2016, 'Values in Transition Design', in G. Kossoff & R. Potter (eds.), *Can Design Catalyse the Great Transition?* [online], pp. 46–55, viewed 11 February 2020, from www.schumachercollege.org.uk/sites/default/files//dissertations/Transi tion_Papers.pdf.

Dahle, C., 2019, 'Designing for transitions: Addressing the problem of global overfish-ing', in T. Irwin & D. DiBella (eds.), *Cuaderno 73: Design in perspective: Transi-tion design monograph*, University of Palermo, Buenos Aires, pp. 213–233, viewed 7 February 2020, from www.academia.edu/36760846/Cuaderno_Journal_73_ Design_in_Perspective_Transition_Design_Monograph.

Debord, G., 2002, 'Perspectives for Alterations in Everyday Life', in B. Highmore (ed.), *The Everyday Life Reader*, pp. 237–245, Routledge, London.

Dreborg, K., 1996, 'Essence of backcasting', *Futures* 28(9), 813–828.

Driscoll, C. & Starik, M., 2004, 'The primordial stakeholder: Advancing the conceptual consideration of stakeholder status for the natural environment', *Journal of Business Ethics* 49(1), 57–73.

Eguren, I.R., 2011, *Theory of Change: A Thinking and Action Approach to Navigate in the Complexity of Social Change Processes.* [online], UNDP and The Hague: Hivos, Panama City, viewed 10 February 2020, from www.democraticdialoguenetwork.org/ app/documents/view/en/1811.

Ellen McArthur Foundation, 2017, *Cities in the Circular Economy: An Initial Explora-tion* [online], Ellen MacArthur Foundation, Cowes, viewed 8 February 2020, from www.ellenmacarthurfoundation.org/publications/cities-in-the-circular-economy-an-initial-exploration.

Escobar, A., 2018, *Designs for Pluriverse: Radical Interdependence, Autonomy, and the Making of Worlds*, pp. 137–164, Duke University Press, Durham, NC.

Gaziulusoy, A.I., 2019, 'Postcards from "the edge": Towards futures of design for sustainability transitions', in T. Irwin & D. DiBella (eds.), *Cuaderno 73: Design in perspective: Transition design monograph*, University of Palermo, Buenos Aires, pp. 67–84, viewed 7 February 2020, fromwww.academia.edu/36760846/Cuaderno_ Journal_73_Design_in_Perspective_Transition_Design_Monograph.

Glaeser, E., 2011, *Triumph of the City*, p. 15, Penguin Books, New York, NY.

Goodwin, B. & Sole, R., 2002, *Signs of Life: How Complexity Pervades Biology*, Basic Books, New York, NY.

Grin, J., Rotmans, J. & Schot, J., 2010, *Transitions to Sustainable Development: New Directions in the Study of Long Term Transformative Change*, Routledge, New York, NY.

Gruber, S. (ed.), 2018, *An Atlas of Commoning: Places of Collective Production*, Institut für Auslandsbeziehungen, Stuttgart.

Guilfoile, C., Tung, K., Urban, M., Cho, E., Sabnis, S. & Yuh, D., 2019, *Decline of Pollinators: Wicked Problem Map*. [blog], viewed 8 February 2020, from https://medium.com/transition-design-decline-of-pollinators/wicked-problem-mapping-2fb294e77f75.

Haagh, L. & Rohregger, B., 2019, *Universal basic income policies and their potential for addressing health inequities*. World Health Organization Report, viewed 22 February 2020, from www.euro.who.int/__data/assets/pdf_file/0008/404387/20190606-h1015-ubi-policies-en.pdf.

Haller, W., 2005, 'Industrial restructuring and urban change in the Pittsburgh region: Developmental, ecological, and socioeconomic tradeoffs', *Ecology and Society* 10(1), 13, viewed 2 February 2020, fromwww.ecologyandsociety.org/vol10/iss1/art13/

Hamilton, S., 2019, 'Words in action: Making and doing transition design in Ojai, California. A case study', in T. Irwin & D. DiBella (eds.), *Cuaderno 73: Design in perspective: Transition design monograph*, University of Palermo, Buenos Aires, pp. 199–2122, viewed 7 February 2020, fromwww.academia.edu/36760846/Cuaderno_Journal_73_Design_in_Perspective_Transition_Design_Monograph.

Hanington, B. & Martin, B., 2019, *The Universal Methods of Design Expanded and Revised*, pp. 230–231, Rockport Publishers, Beverley, MA.

Harvey, D., 2012, 'The Creation of the Urban Commons', In: *Rebel Cities: From the Right to the City to the Urban Revolution*, pp. 67–88, Verso Press, London and New York, NY.

Hiemstra, W., Brouwer, H. & van Vugt, S., 2012, *Power Dynamics in Multi-Stakeholder Processes*, ETC Foundation, Kastanjelaan, viewed 10 February 2020, from https://edepot.wur.nl/242967.

Inayatullah, S., 1998, 'Causal Layered Analysis: Poststructuralism as Method', *Futures* 30(8), 815–829.

Inayatullah, S., 2013, 'Future studies: Theories and methods', in F.G. Junquera (ed.), *There's a Future: Visions for a better world* [online], pp 36–66, BBVA, Madrid, viewed 10 February 2020, from www.bbvaopenmind.com/wp-content/uploads/2013/04/BBVA-OpenMind-Futures-Studies-Theories-and-Methods-Sohail-Inayatullah.pdf.pdf.

Irwin, T., 2011a, 'Design for a Sustainable Future', in J. Hershauer, G. Basile & S. McNall (eds.), in *The Business of Sustainability: Trends, Policies, Practices and Stories of Success*, pp. 41–60, Praeger, Santa Barbara, CA.

Irwin, T., 2011b, 'Wicked Problems and the Relationship Triad', in S. Harding (ed.), *Grow Small, Think Beautiful*, 1st ed., pp. 232–259, Floris, Edinburgh.

Irwin, T., 2015, 'Transition design: A proposal for a new area of design practice, study and research', *Design and Culture* 7(2), 229–246, viewed 7 February 2020, from www.academia.edu/17787817/Design_and_Culture_Journal_Article_Transition_Design_A_Proposal_for_a_New_Area_of_Design_Practice_Study_and_Research.

Irwin, T., 2019, 'The Emerging Transition Design Approach', in E. Resnick (ed.), *Social Design Reader*, pp. 431–451, Bloomsbury, London.

Irwin, T. & DiBella, D. (eds.), 2019b, *Cuaderno 73: Design in perspective: Transition design monograph*, University of Palermo, Buenos Aires, viewed 7 February 2020, from www.academia.edu/36760846/Cuaderno_Journal_73_Design_in_Perspective_Transition_Design_Monograph.

Irwin, T. & Kossoff, G., 2017a, *Mapping Ojai's Water Shortage: The First Workshop*, www.academia.edu/30968737/Mapping_Ojais_Water_Shortage_The_First_Workshop_January_2017.

Irwin, T. & Kossoff, G., 2017b, *Mapping Ojai's Water Shortage: The Second Workshop*, viewed 6 February 2020, from www.academia.edu/32353660/Mapping_Ojais_Water_Shortage_The_Second_Workshop_May_2017.

Irwin, T. & Kossoff, G. (eds.), 2020, *Transition design seminar 2020* [online], viewed 8 February 2020, from https://transitiondesignseminarcmu.net.

Irwin, T., Kossoff, G., Tonkinwise, C. & Scuppelli, P., 2015, *Transition design overview*, Carnegie Mellon School of Design, Pittsburgh, viewed 7 January 2015, from www.academia.edu/13122242/Transition_Design_Overview.

Irwin, T., Kossoff, G. & Willis, A., 2015, *Design Philosophy Papers: Special Issue on Transition Design*. [online], www.academia.edu/20291172/Design_Philosophy_Papers_Special_Issue_on_Transition_Design.

Irwin, T., Tonkinwise, C. & Kossoff, G., 2015, 'Transition Design: An Educational Framework for Advancing the Study and Design of Sustainable Transitions', In *6th International Sustainability Transitions Conference*, Brighton, viewed 8 February 2020, from www.academia.edu/15283122/Transition_Design_An_Educational_Framework_for_Advancing_the_Study_and_Design_of_Sustainable_Transitions_presented_at_the_STRN_conference_2015_Sussex_.

Karavdic, E., Jianxiao Ge, D., Runmaio Shi, M., Ortega Pallanez, M. & Ming-Chieh Chou, M., 2019, *Poor air quality: Multi-level perspective mapping and identifying areas of intervention*, viewed 8 February 2020, fromhttps://medium.com/team-resilience/assignment-2-multi-level-perspective-mapping-af43be406646.

Katz, B. & Nowak, J., 2017, *The New Localism: How Cities Can Thrive in an Age of Populism*, pp. 7–10, Brookings, Washington, DC.

Kostakis, V., Niaros, V., Dafermos, G. & Bauwens, M., 2015, 'Design global, manufacture local: Exploring the contours of an emerging productive model', *Futures* 73, 126–135.

Kossoff, G., 2011, 'Holism and the Reconstitution of Everyday Life: A Framework for Transition to a Sustainable Society', in S. Harding (ed.), *Grow Small, Think Beautiful*, 1st edn., Floris, Edinburgh.

Kossoff, G., 2019a, 'Cosmopolitan localism: The planetary networking of everyday life in place, value', in T. Irwin & D. DiBella (eds.), *Cuaderno 73: Design in perspective: Transition design monograph*, University of Palermo, Buenos Aires, pp. 51–65, viewed 7 February 2020, from www.academia.edu/36760846/Cuaderno_Journal_73_Design_in_Perspective_Transition_Design_Monograph.

Kossoff, G., 2019b, 'Contextualizing Interventions', In *10th International Sustainability Transitions Conference*, Ottawa, viewed 8 February 2020, from www.academia.edu/15403946/Transition_Design_The_Importance_of_Everyday_Life_and_Lifestyles_as_a_Leverage_Point_for_Sustainability_Transitions_presented_at_the_STRN_Conference_2015_Sussex_.

Kossoff, G., Tonkinwise, C. & Irwin, T., 2015, 'Transition Design: The Importance of Everyday Life and Lifestyles as a Leverage Point for Sustainability Transitions', In *6th International Sustainability Transitions Conference*, Brighton, viewed 8 February 2020, from www.academia.edu/15403946/Transition_Design_The_Importance_of_Everyday_Life_and_Lifestyles_as_a_Leverage_Point_for_Sustainability_Transitions_presented_at_the_STRN_Conference_2015_Sussex_.

Kuijer, L., 2014, *Implications of social practice theory for sustainable design*, PhD, [online], viewed 7 February 2020, from www.researchgate.net/publication/266247132_Implica tions_of_Social_Practice_Theory_for_Sustainable_Design.

Kuijer, L. & De Jong, A., 2011, 'Practice Theory and Human-Centered Design: A Sustainable Bathing Example', In *Proceedings Nordic Design Research Conference* (NORDES), Aalto University, Helsinki.

Lockton, D., 2014, *As We May Understand: A Constructionist Approach to 'Behaviour Change' and the Internet of Things*, viewed 2 February 2020, from www.academia. edu/11767037/As_we_may_understand_A_constructionist_approach_to_behaviour_ change_and_the_Internet_of_Things.

Majumder, M., 2015, *Impact of Urbanization on Water Shortage in Face of Climatic Aberrations*, Springer, Berlin.

Manzini, E., 2011, 'SLOC: The emerging scenario of small, open, local, connected', in S. Harding (ed.), *Grow small, think beautiful ideas for a sustainable world from Schumacher College*, pp. 216–231, Floris Books, Edinburgh.

Max-Neef, M., 1991, *Human Scale Development: Conception, Application and Further Reflections*, Apex, New York, NY, viewed 2 February 2020, from www.wtf.tw/ref/maxneef.pdf.

Meadows, D., 1999, *Leverage Points: Places to Intervene in a System*, [online] The Sustainability Institute, Hartfield, VT, viewed February 2020, from www.donellameadows. org/wp-content/userfiles/Leverage_Points.pdf.

Mulder, I. & Loorbach, D., 2016, 'Rethinking design: A critical perspective to embrace societal challenges', in *Can Design Catalyse the Great Transition: Transition Design Symposium* 2016, pp. 16–24, Carnegie Mellon University, Pittsburgh, PA, New Weather Institute and Schumacher College, Totnes, UK.

Niedderer, K., Clune, S. & Ludden, G. (eds.), 2017, *Design for Behaviour Change*, Routledge, Abingdon.

Pohjanpalo, K., 2019, 'How Finland Slashed Homelessness by 40%', *Pittsburgh Post-Gazette*, viewed 6 February 2020, fromwww.post-gazette.com/news/world/2019/07/21/ How-Finland-slashed-homelessness-by-40/stories/201907210139?cid=search.

Porritt, J., 2013, *The World We Made: Alex McKay's Story from 2050*, Phaidon, New York, NY.

Quist, J. & Vergragt, P., 2006, 'Past and Future of Backcasting: The Shift to Stakeholder Participation and a Proposal for a Methodological Framework', In *Futures* 38(9), 10–27–1045.

Rittel, H. & Webber, M., 1973, 'Dilemmas in a general theory of planning', *Policy Sciences* 4(2), 155–169.

Rockefeller Foundation, 2019, *Resilient cities, resilient lives: Learning from the 100RC network*, viewed 8 February 2020, fromhttp://100resilientcities.org/capstone-report/.

Rohrbach, S. & Steenson, M., 2019, 'Transition design: Teaching and learning', in T. Irwin & D. DiBella (eds.), *Cuaderno 73: Design in perspective: Transition design monograph*, pp. 235–263, University of Palermo, Buenos Aires, viewed 7 February 2020, from www.academia.edu/36760846/Cuaderno_Journal_73_Design_in_Perspective_ Transition_Design_Monograph.

RSA Great Recovery Project. 2013, *Investigating the Role of Design in the Circular Economy*, Royal Society of Arts, London, viewed 10 February 2020, from www.thersa.org/ globalassets/images/projects/rsa-the-great-recovery-report_131028.pdf.

Russo, J., 2017, 'The Pittsburgh Conundrum: Can You Have a Model City in a Left Behind Region', *The American Prospect*, viewed 10 February 2020, from https://prospect.org/ labor/pittsburgh-conundrum/.

Seed, J. & Macy, J., 2007, *Thinking Like a Mountain: Towards a Council of All Beings. New Catalyst Books* New Catalyst Books, Gabriola Island.

Shove, E., Pantzar, M. & Watson, M., 2012, *The Dynamics of Social Practice: Everyday Life and How it Changes*, Sage Publications, London, UK.

Shove, E. & Walker, G., 2010, 'Governing transitions in the sustainability of everyday life', *Research Policy* 39, 471–476.

Simon, M. & Rychard, S., 2005, *Conflict analysis tools*, retrieved from the Swiss Agency for Development and Cooperation (SDC) website, https://css.ethz.ch/content/dam/ethz/special-interest/gess/cis/center-for-securities-studies/pdfs/Conflict-Analysis-Tools.pdf.

Singh, D., Zheng, C., Ploehn, C. & Khoshoo, A., 2019, *Assignment 2: Multi-Level Perspective Mapping*, viewed 8 February 2020, fromhttps://medium.com/lack-of-access-to-healthy-food/multi-level-perspective-mapping-c6155cda3372.

Thackara, J., 2015, 'Commoning: From Social Money to the Art of Hosting and Knowing', In *How to Thrive in the New Economy: Designing Tomorrow's World Today*, pp 135–168, Thames and Hudson, New York, NY.

Toderian, B., 2015, 'Better city-making means breaking down silos – Here's how', *Planetizen*, viewed 7 February 2020, from www.planetizen.com/node/80172/better-city-making-means-breaking-down-silos-heres-how.

Tonkinwise, C., 2015a, *Design for transition – From what and to what?* [online], viewed 7 February 2020, from www.academia.edu/11796491/Design_for_Transition_-_from_and_to_what.

Tonkinwise, C., 2015b, *Crafting transition designs: The urgency of the slow's resistance to the big. Making futures journal*, Plymouth College of Art, Plymouth, viewed 3 February 2020, fromhttps://drive.google.com/file/d/1qvdQV-DTHZOrW8hKZrzq63CbJyg6Bzt/view.

Tonkinwise, C., 2018, 'Concerning Relations in the City: Designing Relational Services in Sharing Economies', in L. Vaughan (ed.), *Designing Cultures of Care*, pp. 189–202, Bloomsbury, London.

Tonkinwise, C., 2019, 'Design's (dis)orders: Mediating systems-level transition design', in T. Irwin & D. DiBella (eds.), *Cuaderno 73: Design in perspective: Transition design monograph*, University of Palermo, Buenos Aires, pp. 85–95, viewed 7 February 2020 from www.academia.edu/36760846/Cuaderno_Journal_73_Design_in_Perspective_Transition_Design_Monograph.

Trinko, K., 2020, Tents, Homelesssness, and Miserty: 9 Things I Saw in San Francisco, [online] *The Daily Signal*, viewed 8 February 2020, fromwww.dailysignal.com/2020/01/13/tents-homeless-and-misery-9-things-i-saw-in-san-francisco/.

Trist, E., Emery, F. & Murray, H. (eds.), 1997, *The Social Engagement of Social Science, vol. 3: The Socio-Ecological Perspective* University of Pennsylvania Press, Philadelphia.

Trist, E. & Murray, H. (eds.), 1993, *The Social Engagement of Social Science, vol. 3: The Socio-Technical Perspective*, University of Pennsylvania Press, Philadelphia.

Wheatley, M., 2006, *Leadership and the New Science: Discovering Order in a Chaotic World*, Berrett-Koehler, San Francisco.

6 Transition pioneers

Cultural currents and social movements of our time that "preveal" the future post-capitalist city

Juliana Birnbaum

Introduction

> *Thinking has come full circle on cities, from blaming them for environmental destruction to considering that urban environments, properly designed and managed, can be a kind of biological as well as a cultural ark.*
> — Paul Hawken (2017)

The next-generation sustainable city as ark, carrying ancient and cutting-edge technologies through the sea of our ecological crisis, is a powerful vision. A post-capitalist city, in simple, permaculture-inspired terms, could be seen as one that has broken out of current global systems that are harming people and planet, systems that are creating growing inequality. It is a resilient urban community with local food and energy sources, zero waste, participatory democracy and a culture that promotes care for people and egalitarian access to resources.

Human habitats are becoming increasingly urban as our global population grows to what some predict will be its maximum number by the middle of this century. By 2050, two-thirds of the world's population (an estimated 6.7 billion people) will live in urban environments, according to a 2018 report (United Nations 2018). As cities continue to grow over the coming decades, they will become a key location for the redesign of our systems on multiple levels that is necessary for the health of civilization in the face of population growth and the climate crisis. A swift transition away from fossil fuels and the systems that waste them is essential to drawing down atmospheric carbon levels, and from where I'm writing in California in 2019, that seems to be clear to most governments in the world except my own, overtaken by the entrenched corporate capitalist interests embodied by the Trump administration.

The post-capitalist city must be designed or retrofitted to put ecology first, with an approach that considers the interwoven whole of our connections with each other, as well as our broader relationship with the natural world. This vision of a city is tied in with a much more comprehensive, global cultural response to climate change and the failure of capitalism as a global system to maintain a healthy planet and culture. The many aspects of the movement that make up this shift draw on knowledge that is ancient, yet integrated with modern science and technology. It must involve the unwinding of our collective past and a reckoning with colonial histories of indigenous genocide and enslavement, along with a re-visioning of a way forward that offers an alternative imagination, one where life's interconnectivity is writ large.

> The first step towards reimagining a world gone terribly wrong would be to stop the annihilation of those who have a different imagination – an imagination that is outside of capitalism as well as communism. An imagination which has an altogether different understanding of what constitutes happiness and fulfillment. To gain this philosophical space, it is necessary to concede some physical space for the survival of those who look like the keepers of our past, but who may really be the guides to our future.
>
> (Arundhati Roy 2010)

The current global ecological crisis is being forced by the "invisible" hand of dominant neoliberal capitalism – now made painfully visible. It calls for a collective cultural cognition and reaction (or recognition and action if you prefer). This response, which has been compared to the interconnected activation of an organism's immune system when the health and integrity of the body is threatened, *is already happening.*

The movement is taking shape in the form of a multitude of fractal "nodes," or centers of human activity. Together these separate nodes become a broader network that constellates this growing cultural response, revealing a blueprint for building or retrofitting resilient cities that generate, use, share and dispose of resources wisely. In sum, the present-day initiatives described here demonstrate the collective emergence of change on a global scale and also a vision for a post-capitalist city, one in which the linear flow of materials from factory to dump has been converted into a closed-loop, sustainable cycle.

This chapter will focus on present-day, alternative international movements related to three major sectors – food, energy and housing – and profile communities that are revealing the lines and shapes of our potential city of the future. It draws on research undertaken as part of the Sustainable [R]evolution project I co-founded in 2006, where I and a group of researchers studied and documented nearly 100 regenerative design sites. It is also informed by research I and numerous others did as part of Project Drawdown, a coalition of scholars and scientists from across the globe who are mapping and measuring a collective array of substantive solutions to global warming with the goal of reaching drawdown: the point in time when the concentration of greenhouse gases in the Earth's atmosphere begins to decline on a year-to-year basis.

First, we look at present-day movements to **relocalize food** in urban centers large and small to incorporate inner-city and regional farms, gardens and woodlands. This patchwork (ideally network) of green helps to restore urban ecosystems and positively impact a city's carbon footprint, while promoting food security and connecting people with healthy local food. In and around urban areas, food is being grown, using both low-tech methods in vacant lots and with sophisticated practices of rooftop and vertical farming with hydroponic and other technologies.

Second, we discuss cities developing **decentralized renewable energy** run on localized smart grids, examining places that have made or are making this transition. This section will mention some of the renewable power sources that can be tapped into when this transition is made (wind, solar, district heating, landfill methane and geothermal). This shift in how cities *get* energy goes hand in hand with a shift in how they *use* that energy, so this section discusses district heating/cooling and net zero architecture with its associated technologies and briefly mentions sustainable transportation systems as well.

Finally, perhaps the most profound part of this work involves a re-imagining of the social aspect of human settlements along with the ecological aspect. Work in these realms is visible in today's **cohousing** movement, and we will examine the models present in those intentional communities that prioritize environmental sustainability, often known as ecovillages. We will discuss some of these projects located within urban areas in a diverse range of ecosystems and climate zones. These global cultural currents, supported by numerous others, together begin to "preveal" the communal, resilient, even regenerative urban settlement of tomorrow. That is to say: these movements manage to reveal the face of that future city *before* it has actualized.

Decentralizing control over energy, water and food, along with reducing waste, are essential to any community, from village to megacity, with the goal of climate resilience. In *This Changes Everything: Capitalism vs. the Climate*, her 2015 book, Naomi Klein writes (emphasis added):

> The relationship between power decentralization and successful climate action points to how *the planning required by this moment differs markedly from the more centralized versions of the past.* There is a reason, after all, why it was so easy for the right to vilify state enterprises and national planning: many state-owned companies were bureaucratic, cumbersome, and unresponsive; the five-year plans cooked up under state socialist governments were indeed top-down and remote, utterly disconnected from local needs and experiences. . . . The climate planning we need is of a different sort entirely. . . . *Communities should be given new tools and powers to design the methods that work best for them* – much as worker-run co-ops have the capacity to play a huge role in an industrial transformation.
>
> (2015:133).

The next green revolution will happen in the city: embedding regenerative agriculture into and around urban centers

The 20th-century version of the "Green Revolution" for agriculture attempted to address the need to provide food for an exponentially growing population through genetic engineering, intense application of chemical fertilizer and a general

industrialization of farming heavily reliant upon fossil fuels. This approach provided some short-term results at the cost of the long-term sustainability and health
of people, regional food systems and, as it turned out, the planet. Today our food-
related emissions are the number one cause of global warming. This includes the
impacts of deforestation and methane pollution related to raising animals, and
the heavy carbon footprint of agricultural systems that include powering heavy
machinery, transport, packaging, refrigeration of goods and finally the food waste
at the end of the line.

In response to this reality, the new millennium has brought a significant trend
toward techniques and practices that have the ability to transform food production
from a destructive force to a means for sequestering carbon in the soil and drawing
down emissions. The autonomous city of the future must be built upon a secure
and sustainable food system, with local, small-scale farms embedded within or in
close proximity to the urban center, and systems for composting organic waste and
closing the loop by applying it back to soils. Urban farms can restore degraded
land with regenerative practices including the use of diverse cover crops, avoiding
tillage which wastes water and releases the carbon in soil into the atmosphere, and
multistrata agroforestry or tree intercropping where farmers plant trees among
crops to create a layered food forest.

By necessity, the island nation of Cuba became an innovator of re-localized and
sustainable food systems from the collapse of the Soviet Union in the late 1980s.
The country's food supply went into a dangerous free fall, bringing Cubans to
the brink. Somehow the country needed to produce far more of its own food with
far fewer outside inputs, or it would face widespread starvation and the end of its
political and economic sovereignty.

Drawing on a well-educated and resourceful population, Cuba's response
to the crisis was a wide-ranging package of land-reform measures and agroecological farming methods that significantly altered the country's agricultural
landscape within a few short years. Some of the key initial features of this
transformation were the establishment of large numbers of family farms and
cooperatives, strategies to increase biodiversity and soil fertility, the revival of
the use of animal traction, and widespread composting operations, including
vermicomposting.

In an effort to ramp up food production in close proximity to where the majority of the population lived, in the early 1990s the Cuban state mandated com-
mercialized urban organic agriculture, called organoponico, in towns and cities
throughout the country (Birnbaum & Fox 2014:55). Today, over 300,000 urban
farms produce about 50% of the island's fresh produce, along with 39,000 tons of
meat and 216 million eggs.

Besides providing local food security and reducing the significant environmental costs of shipping and storing food grown long distances from where it is con-
sumed, urban agriculture can solve other problems at the same time. Known as
"green infrastructure," the practice of micro-farming within cities – utilizing rooftops and vertical space – cools and filters the air and allows rain to be absorbed

and held in the soil, instead of causing storm water runoff issues such as flooding and pollution. Other benefits include carbon sequestration, insulation (which reduces heating/cooling costs), improved air quality and rainwater storage.

Singapore has become a hot spot for the development of both green rooftops and vertical gardens, due to its commitments to green standards and a proactive economic stimulus system where installation is half covered by the government. The city also has mandated sustainable building codes (National Geographic 2017). In Hong Kong, the past decade has seen the emergence of hundreds of rooftop and vertical farms using state of the art technology for growing food. Some of these include hydroponics, aeroponics, aquaponics, modular and cubic farming systems (Tsui 2019).

In the United States, urban farming has increased by more than 30% in the past three decades (The Conversation 2019). New York City has become a leader within the United States for rooftop farming, supplying its population with locally grown and organic produce. Its largest rooftop farm as of 2019 is in Brooklyn and its storm water runoff management capability allowed it to be supported by government Green Infrastructure grants. The NYC Urban Planning Department has updated its zoning regulations to encourage green development and rooftop farms (Foderaro 2012). In 2018, the USDA and the Department of Energy held a conference focused on vertical agriculture and sustainable urban ecosystems, signaling a movement toward prioritizing these innovations on a government planning level (USDA 2019).

In the Midwest, the effects of globalization on industry has turned some of the former grand manufacturing hubs of the United States into sprawling patchworks of vacant lots and abandoned buildings. During an economic boom, urban agriculture is typically threatened, during a bust, vacant land is plentiful and there is limited demand, so urban farms flourish. Detroit, Michigan has become one example of a city where decaying lots (abandoned due to changes in the auto manufacturing industry), have been broadly transformed into urban woodlands and farms.

Integrating agriculture into capitalist cities in the present day – especially those affected by urban decay – has not been without its challenges. A significant part of Detroit's transformation has been facilitated by what some call a "land grab" on the part of John Hantz, who purchased a large number of vacant lots in order to plant trees and bring up property values – whether he is a hero or a villain depends on who you are talking to. Authors of a study of Philadelphia and the growth of urban farms there write that "Ongoing tensions surround the role played by urban agriculture in . . . a city that after decades of economic depression and abandonment has begun to attract new development attention and gentrification" (Rosan & Pearsall 2017:6–7).

Despite the growing pains of the current movement, it is clear that integrating agriculture into and around our urban centers is an essential part of transforming current systems contributing to climate change and developing a resilient post-capitalist culture.

Smart grids deliver renewable energy to net-zero buildings and neighborhoods

The regenerative city of the future needs energy resilience. It relies on decentralized energy grids, running on renewable energy sources such as wind, solar, landfill methane and geothermal. Its micro or "smart" grids are capable of storing and moving energy with maximum efficiency. As cities transition their energy sources, they also need to shift the way that energy is used. The post-capitalist city makes use of the appropriate technologies for buildings that have the goal of having a net zero energy consumption. It uses strategies such as district heating and cooling systems, highly efficient economies of scale that remove the responsibility of maintaining these systems from the individual (Johnson 2014). Along with these shifts in the way that energy is sourced and used, a major transition toward sustainable public transport systems and walkable/bikeable urban environments must take place as part of any broad transition plan.

Microgrids can connect to (or disconnect from) larger power grids and enable communities from small and remote to highly urban to be more self-reliant. They provide energy resilience and can offer recovery options for regions experiencing extreme weather events. They are localized, independent energy grids that can draw upon various energy sources, tailored to the most sustainable options for the region. West Marin County, where I live on the rural edge of the San Francisco Bay Area, is considering microgrids to create energy resilience in the face of the increasing wildfire threat in California – many of the highly destructive wildfires in recent years here were caused by power grid infrastructure (Houston 2019). Days that are red flagged for fire danger could be reason for energy providers to turn off the grid to reduce risk. As we head into fire season this year, we have been warned repeatedly that our energy may be turned off for days at a time. Microgrids essentially create islands of power that can be tapped in emergency situations, or adapted to provide year-round energy for a village or an entire city.

In 2012, following the earthquake and tsunami-triggered Fukushima Daiichi nuclear power plant meltdown, the Japanese government established the "Future City" initiative to develop solutions for dealing with environmental challenges, climate change and disaster preparedness. They chose the city of Higashimatsushima, which suffered the worst flood damage in the region in the aftermath of the disaster, leaving 65% of the city underwater, killing 1,100 people and displacing about 10,000. The goal was to make Higashimatsushima a "net zero energy city" by 2022, running on locally produced energy. Japan's first microgrid community was built there, running on photovoltaic panels and a bio-diesel generator. The entire city will be powered by renewables and have its own large-scale energy storage (Cohen 2018:160–163).

New York University, in lower Manhattan, has produced power onsite since 1960, and in 2008 began constructing a cogeneration plant producing both electricity and steam which is used for district heating. During blackouts, NYU has been an island of power in a dark city, and has enabled the university to reduce its CO_2 emissions by more than 30% (Cohen 2018:154).

The future sustainable city has learned how to maximize heating and cooling efficiency through collective district systems. These systems consist of an infrastructure of pipes linking consumers to neighborhood thermal stores. A large district heating network operates in Pimlico, a district of London, built in the 1950s to supply 1,600 social housing units with heat produced across the river in Battersea Power Station. While the coal-fired plant closed decades ago, the heating infrastructure was resilient and today this neighborhood emits 8% less CO_2 than if everyone was using their own boiler (Johnson 2014).

Copenhagen made it mandatory in the 1970s that urban inhabitants had to connect to city systems for their heat, to meet their goals of minimizing waste and cutting national energy imports. Today the city has one of the most impressive district heating networks in the world: 98% of the city's heating needs are met through heat networks, and these are now being used to deliver Denmark's commitment to fossil-free heating and electricity by 2035.

"The Danish model is celebrated as proof that once cities have pipes installed, they can gradually change what is connected to either end," reported *The Guardian* in 2014. "This could be superefficient, low-energy buildings that don't need much heat, or carbon-neutral energy sources such as geothermal, or innovations to store 'surplus wind' as thermal energy" (Johnson 2014).

District cooling is becoming more important as cities experience more severe and common heat waves on our warming planet. It can work by providing chilled water (from the sea or other sources) to buildings for cooling, or via a circuit or Heat Sharing Network using a heat pump to release heat into the ground. Systems such as these are in use in Helsinki, Munich, Geneva, Ontario, Abu Dhabi and other cities around the world. In India, the Gujarat International Finance-Tec (GIFT) City is taking credit for the country's first district cooling system. The city is a Special Economic Zone designed to be a financial and IT services hub and is powered from two different substations to provide energy resilience (REHVA 2018).

While the GIFT City system may be the first on its scale in India, an ecovillage in Tamil Nadu called Auroville actually pioneered a district cooling system decades ago along with numerous other ecological innovations. Its Earth Cooling Tunnel pulls hot air through a pipe that runs through the bottom of a massive rainwater harvesting tank, allowing it to cool by heat exchange before the air is pumped into residential units. Founded in 1968, Auroville is known as the most eco-friendly and pollution-free city in India and is home to a large photovoltaic power plant, over 30 windmills, 20 biogas units, and a unique solar concentrating technology known as the "solar bowl." It uses one of the largest reflecting bowls in the world and a heliostat to track and focus the Sun's rays to power the community's enormous kitchen.

Cities can tap into the renewable energy source most available and best suited for their particular climate and region, from wind (estimated to be the cheapest energy source available in most economies) to solar and geothermal. Circular systems that make use of neighborhood waste to provide energy are ideal, such as landfill methane systems and biodigesters.

One of the benefits of biodigestion as a technique for producing energy is that it is just such a closed-loop system, where wastewater or manure can be treated safely and turned into an asset rather than left to contribute to the greenhouse gases in the atmosphere. Similar technologies have been used in urban settings to treat wastewater sustainably and effectively, while at the same time creating beautiful spaces within cities. The Baima Canal in Fuzhou, China uses tens of thousands of native plants to treat its industrial wastewater and sewage, which was formerly dumped untreated into the Minjiang River. Today the waste is run through a "linear restorer" biological treatment system that has transformed a polluted canal into a beautiful neighborhood common space with a pedestrian walkway (Birnbaum & Fox 2014:159). The project shows the intersection of environmental and social change that such systems can promote, which we return to later in our dive into the ecovillage movement.

Net zero architecture in the post capitalist city

Cities on the path to zero net energy consumption are changing not only the sources of their power, but are building to ensure that it is used in the most efficient way possible. Buildings in the United States contribute more emissions than every vehicle combined when considering the energy used to build, heat, cool and run appliances (Roudman 2013). A net-zero building is one that has zero net energy consumption, producing as much energy as it uses in a year. In some months it may generate excess electricity, at other times it may require electricity. There are many variations on the design of such buildings, ranging from Earthships, which originated in New Mexico and use recycled tires as insulation, to high-profile urban office buildings in India, China and Malaysia. Net-zero buildings employ a broad range of site-specific sustainable technologies including the green roofs described earlier.

There are hundreds of green building certifications and rating systems worldwide, and some (including the most popular, LEED) have been criticized for rewarding minor, low-cost steps that have limited environmental benefit. The U.S. Green Building Council and its LEED certification has aided developers in winning tax breaks, incentives and grants while charging higher rents, while some of its most high-profile buildings have failed to be any more energy efficient than those of conventional construction. Designers can target the cheapest and easiest "green points" options rather than more costly cutting-edge technologies with significant results. Certification is awarded before occupancy, meaning that points are based on projections rather than actual data on energy and water savings (USA Today 2012). That being said, I believe that some of these standards can effectively promote net-zero buildings, and I'll focus here on one certification which provides a good example of their potentially positive role in the emerging direction of urban design.

Passive House is an internationally recognized, stringent energy performance standard for buildings that can be a path to net zero and is one of the fastest-growing metrics of its kind worldwide. It is a design methodology where energy losses are minimized and passive heat sources are maximized. An approach with proven success for constructing comfortable, ultra-low – energy buildings in a variety of climates and cultures, it rests mainly on insulation, airtightness and ventilation. Structures are warmed by sources such as the Sun, appliances and heat recovery and use strategic shading in hotter months to stay cool.

The *passive* in "passive house" refers to reducing a building's heating and cooling demand by 90–95% with minimal use of expensive, non-renewable *active* technologies such as conventional fuel/electric systems or even photovoltaics. The concept allows a comfortable indoor temperature without the need for an active heating or cooling system. Its buildings must have a ventilation system that circulates fresh air, but due to the high thermal quality of the building and airtightness, the house needs only to be heated by passive heat sources, such as sunlight, body heat and appliances. When compared with homes designed along conventional building codes, Passive Houses use dramatically less energy overall (from 50% up to 80%) and provide greater thermal comfort and excellent indoor air quality (Boer & Kaan 2006).

The standard and others similar to it are focused on lowering energy consumption first and are flexible according to site, allowing for creativity and the use of locally available and adapted building materials. In the past decade, net-zero buildings and neighborhoods have started to become commonplace and some strive to be energy positive, such as the Sonnenschiff solar city in Freiburg, Germany, which produces four times the energy it consumes (Hawken 2017:84). Newer net-zero buildings push the margins further with goals of zero water and zero waste, harvesting rain and processing sewage onsite into compostable forms. In California, the strategic plan calls for all new residential construction to be zero net energy by 2020, and all commercial construction by 2030.

The transition from outmoded 20th-century materials systems to renewably sourced microgrids feeding net zero energy and waste neighborhoods is in motion, but needs to be taken up much faster. It requires the collaboration of government agencies with communities, energy companies and other stakeholders, toward investment, research and development (Cohen 2018:153). In the United States, the "Green New Deal" is being proposed by some Democratic representatives and presidential candidates, which would aim to address the country's growing economic inequality and the climate crisis through a set of social and economic reforms and public works projects. It references the initiative undertaken by President Roosevelt in the 1930s as a response to the Great Depression that successfully reinvigorated the US economy, and would involve transitioning the energy infrastructure to renewables on a large scale. The Green New Deal has been popularized in Britain and has also been taken up on an international level by the United Nations.

The ecovillage as a social and environmental force

Thousands of models exist for the development of sustainable and resilient food, energy, building and transportation systems necessary for the post-capitalist city, some of which were discussed earlier. Yet, another related but distinct movement is growing exponentially, adaptable to scale and climate. It is the global ecovillage movement, made up of communities on the growing edge of cultural and social innovation, as well as environmental. Often planned, managed, and owned by residents, ecovillage communities are typically made up of private living spaces (apartments or houses) and shared facilities such as offices, communal kitchen, laundry, gym and other facilities, classrooms/playrooms, and gardens or farmland. They are designed to promote intergenerational community by maximizing social interaction between residents, who often share meals and other activities and park cars on the periphery of the development to create common pedestrian spaces.

Ukrainian sustainability consultant Vitaliy Soliviy writes:

> What separates ecovillages from cities is not merely the size or density of the population. Rather it's the philosophy of sustainable living and cultural transformation that comes before innovation and speed. These are places where joy from manual labor tops efficiency, while the focus on consciousness and connection to nature is not just about daily relief from a busy life. It's about actually finding a way to live within nature and not outside it.
>
> (Sustainability Times 2019)

In my years of researching intentional communities with the Sustainable [R] evolution project, again and again the people I spoke to said something like this: "Figuring out the technology involved in building green or switching to renewable energy is the easy part. The challenge is learning to live together in community and re-imagining the social aspect of human settlements."

Those who are pioneering the cohousing and ecovillage movements in today's world face the growing pains involved in bringing an intentional community together within a capitalist culture that emphasizes independence and the isolated, self-reliant nuclear family, rather than interdependence within a larger neighborhood or community group. Despite these hurdles, the ecovillage movement is growing worldwide and starting to have an important influence on mainstream urban planning groups, such as the U.S. Department of Housing and Urban Development. In 2019, U.S. Housing Secretary Ben Carson visited the Hawthorne EcoVillage in Minneapolis, Minnesota, a newly constructed 75-unit affordable housing community, touring a rooftop where solar panels and rainwater catchment systems are installed. Carson said he'd like to see more local governments follow the example of Minneapolis, which has eliminated single-family zoning in an effort to address homelessness (Evans 2019).

According to the Global Ecovillage Network, over 10,000 ecovillages exist today (Global Ecovillage Network 2016). Hundreds exist in Denmark and northern Europe, with about 300 in the Netherlands, 165 in the United States and about 140 more in the planning stages (There's Community and Consensus: But It's No Commune 2018) and many more in Canada, the United Kingdom, Australia, Latin America, Asia, the Middle East and Africa. While these communities can be rural, suburban or urban, they often cluster the housing to allow for more shared facilities and social interaction, leaving the remaining land for shared use and wildlife conservation. Ecovillages are cohousing communities that have prioritized environmental sustainability in their design and use of resources. Innumerable traditional villages, indigenous tribes and religious communities can be seen as ecovillage prototypes. Their contemporary counterparts started with the "back-to-land" movement in the 1960s and draw on earlier waves of intentional community movements that occurred in many countries. In Denmark, the cohousing movement began in 1968, inspired by a newspaper article related to the social benefits of communal life called "Children Should Have One Hundred Parents."

Jonathan Dawson, former president of the Global Ecovillage Network, describes five ecovillage principles:

1　They are not government-sponsored projects, but grassroots initiatives.
2　Their residents value and practice community living.
3　Their residents are not overly dependent on government, corporate or other centralized sources for water, food, shelter, power and other basic necessities. Rather, they attempt to provide these resources themselves.
4　Their residents have a strong sense of shared values, often characterized in spiritual terms.
5　They often serve as research and demonstration sites, offering educational experiences for others (Taggart 2009:20).

Ecovillages can have social impact as well as environmental, creating spaces for democratic process and the empowerment of marginalized people. At Nashira, an ecovillage near the city of Cali, Colombia, a group of women (many of them single mothers widowed by decades of conflict, or survivors of domestic or sexual abuse) founded and constructed a community based on ecological and social principles. Nashira is 88-year-old, low-income woman, head-of-households, whose mission is to go "beyond offering just housing solutions," according to its literature.

> Nashira seeks to provide a better quality of life, offering a secure and nutritious supply of food within the compound, an environmentally friendly atmosphere and a source of income through the development of workshops where women can manufacture their own products.
>
> (Permaculture Magazine 2013)

Residents grow their own food and produce income selling the surplus at market, and offer educational workshops on green building, participatory decision-making, renewable energy and other topics related to the ecovillage (Marlens 2019).

The ecovillage model can be scaled to urban centers, and it can also be replicated in smaller communities and used to retrofit existing towns. In 2015, the first International Ecovillage Summit was held in the city of Dakar, in Senegal. Its government is using ecovillage principles to transform communities throughout the country. Hundreds of villages have already been equipped with ecovillage techniques, with the goal being to transform every second village of the country into an ecovillage (Dregger 2015).

At the time of writing, in 2019, Danish architects are in the process of designing a large, next-generation ecovillage and construction is to begin soon. The designers aim to use all 17 of the United Nations Global Sustainability goals as a blueprint for a 400-home ecovillage on the southern outskirts of Copenhagen that will set a new standard for sustainable building. UN17 Village will house 830 people, including around 175 children and 100 older residents. Five housing blocks will be built using recycled concrete, wood and glass (unenvironment. org 2019).

This final section goes into more detail on two well-established urban ecovillages in the United States, located in two very different climate zones. I visited both the Los Angeles Ecovillage in California and the EcoVillage at Ithaca in New York in researching *Sustainable [R]evolution: Permaculture in Ecovillages, Urban Farms and Communities Worldwide*, published in 2014, which enlisted a small army of contributors to research and document community-based permaculture sites around the globe. While both of these are still firmly embedded within capitalist structures in the United States, they also present a challenge to globalized, corporate consumer-driven economies and demonstrate change happening from within current systems.

Los Angeles Eco-Village, California

Just off one of the most congested traffic corridors in Los Angeles, tiled with a mosaic of fast-food chains, nail salons and dollar stores, lies a little green oasis: the Los Angeles Eco-Village (LAEV). Three miles west of downtown LA, it is a place of urban regeneration and innovation in the warm Mediterranean climate of southern California. Established in 1996, the ecovillage includes about 50 apartments and acts as a central hub for many green activities and campaigns in the city.

Two blocks of 1920s apartment buildings have been retrofitted through a non-profit organization guided by a permaculture approach to problem solving. Within its grounds, LAEV has facilitated technology and lifestyle changes, such as installing solar panels and composting facilities, providing rent reductions for people who live car-free. It has transformed its courtyard into a 7,000 square foot (650 square meter) garden that provides fruits, herbs and vegetables as well as a lush common area to sit and relax in. LAEV has also incubated businesses like the Bicycle Kitchen – a shop that repairs bikes and that trains neighborhood children in bicycle maintenance skills.

The community has influenced the broader political process of Los Angeles, from lending support to Green candidates to engaging in public planning processes, such as the restoration of the Los Angeles River, and local redevelopment – all while continuing to be an affordable, transportation-accessible place to live.

Despite being in a highly urban space, LAEV has placed a particular emphasis on nature stewardship and in expanding local food sources. The blocks stand out from their surroundings in the amount and variety of greenery, with fruit trees and flowers edging out into the road. The courtyard is a forest of edible plants around a huge magnolia tree with tomatoes, herbs, peppers, chard, borage, bananas, peaches, apples, apricots, figs, mint and comfrey. A flock of chickens are integrated into the garden system, and bees, housed on the roof, as well.

Arkin explained, "We have a foot of mulch in the courtyard – we call the soil our future, our wealth." The courtyard not only uses every inch of space to grow food, but facilitates interaction between neighbors and consequently communal responsibility for looking after, sharing and enjoying the space. Hidden nooks, chairs and hammocks are shaded by fruit trees, and colorful murals bring life to the walls. The outside space is just as lived in as indoors. This has extended beyond the formal boundaries of the village. Members recently secured permission to grow macadamia nut trees along the sidewalk (after long health and safety discussions about people slipping on fallen fruit) and are in the process of establishing a learning garden across the road where they will run workshops for school children.

Ecovillage residents are also striving to make use of every inch of space for slowing and storing water. "We broke up the concrete and created the first permeable sidewalk in L.A., mulched the pathways of our courtyard, and are harvesting rainwater from roof feeds," Arkin recounted. "We cooperated with a local youth center to create an eco-park with a stream fed by a storm drain, and a gravel bed that allows storm water to return to the water table." Stacking multiple solutions, LA Eco-Village demonstrates the potential for urban communities to retrofit city blocks and transform a city's culture from within.

EcoVillage at Ithaca, New York

Founded by two single mothers, Joan Bokaer and Liz Walker, in 1991, EcoVillage at Ithaca (EVI) was the first cohousing community to be established in the eastern United States. It includes three neighborhoods on 176 acres (71 hectares) where a developer had initially planned 150 conventional subdivision homes. The first two neighborhoods had 30 attached homes each, the latest, at the time of this writing, developing 40 units, creating a resident population of about 500. In alignment with permaculture principles, the land is zoned in a way that accounts for both human and natural patterns of use. About 90% of it is green space in the form of deciduous forest and farms; and 10% is developed. Ithaca, located two miles from the ecovillage, is a medium-sized city that is home to several universities, with the hot summers and cold, snowy winters typical of a continental climate.

"The ultimate goal of EVI is to redesign human habitat by creating a model community that will exemplify sustainable systems of living – systems that are not only practical in themselves but replicable by others," reads the ecovillage mission statement (Ecovillageithaca.org 2020). The project demonstrates the

Figure 6.1 Climate Strike. The author's daughter, Lila Fox, at the Global Climate Strike on September 20, 2019, considered the largest climate action day in history with over four million participants worldwide. Her sign with the watchful eyes reads "WAKE UP NOW!" Photograph printed with permission by Louis Fox.

Figure 6.2 Climate Strike. The September 2019 climate strikes were part of the Global
Week for Future, a series of international protests to demand action to address
the world's climate crisis. Images of the actions flooded social media. Lila
Fox captures a photo of her friend marching with her at the rally on Market
Street, San Francisco, California. Photograph printed with permission by
Louis Fox.

feasibility of a site design that provides housing, energy, social interaction and
food production while supporting natural ecosystems. The ecovillage includes
a community-supported agriculture plot that feeds about 1000 subscribers in
the growing season and a "pick your own" berry farm with five varieties of
berries.

On average, households at the ecovillage use about half of the energy and half
of the water of an average American house. A number of homes have rainwater
catchment and/or greywater systems, and a photovoltaic array on the rooftops
meets about half of the community's energy requirements.

The ecovillage is designed to foster intergenerational community and is
designed with spaces that invite sharing and connection between members both
young and elder. Having the central corridor between homes car-free, so that front
yards open up onto communal space, allows for neighbors to come into more contact. It also makes play much safer for children, who can explore the indoor and
outdoor spaces of the ecovillage.

Figure 6.3 Cuba. The Indio Hatuey Research Station in Matanzas, Cuba, focuses on innovating new agroecological methods as part of the state-sponsored urban agriculture program. Photograph by Ron Berezan, from Sustainable Revolution: Permaculture in Ecovillages, Urban Farms, and Communities Worldwide published by North Atlantic Books, copyright © 2014 by Juliana Birnbaum and Louis Fox. Reprinted by permission of North Atlantic Books.

Figure 6.4 EARTHSHIP. In this Earthship in Taos, New Mexico, the greywater from the kitchen sink is filtered and irrigates raised beds. Once filtered again, the water will be used to flush toilets before going into a blackwater treatment planter outdoors. Photograph by Louis Fox from Sustainable Revolution: Permaculture in Ecovillages, Urban Farms, and Communities Worldwide published by North Atlantic Books, copyright © 2014 by Juliana Birnbaum and Louis Fox. Reprinted by permission of North Atlantic Books.

Walker writes:

> Child-rearing in our cohousing neighborhood differs from the mainstream. Most kids seldom if ever watch TV – they are far too busy participating in the day-to-day activities of the village community. At EVI children learn conflict resolution skills from the earliest age. And the six to twelve year olds participate in a "Kids Council" where they take part in developing their own behavioral guidelines and doing fun projects, such as building a bike trail.
>
> (2005)

Marty Hiller, part of Land Partnership Committee that works with permaculture design at the ecovillage (EVI), said:

> One of the things I like about having a kid here is that she gets to go out to the chicken house to gather the eggs, or climb a cherry tree and eat cherries.

Figure 6.5 EARTHSHIP. Earthships are designed to collect and store their own energy, mainly through passive and active solar technologies and wind power. Photograph by Louis Fox from Sustainable Revolution: Permaculture in Ecovillages, Urban Farms, and Communities Worldwide published by North Atlantic Books, copyright © 2014 by Juliana Birnbaum and Louis Fox. Reprinted by permission of North Atlantic Books.

These are shared resources, so I don't have to do all of the work myself to maintain those things.

It's a much more social environment than a regular suburb. She has taken a wild edibles class, she can poke around in the woods and find food. She's growing up able to run around wild, with a really nice set of experiences, not just supervised playdates.

(Hiller 2008)

Community meals are offered nearly every other day in one of the common houses, which house communal kitchen and workshop spaces and also include offices, a playroom and shared laundry facilities. The EcoVillage at Ithaca was the first of its kind cohousing community on the East Coast of the United States and is now one of hundreds.

Figure 6.6 EARTHSHIP. An Earthship owner maintains a water catchment gully on her roof. Her home is designed to catch and store enough water from rain and snow to meet the needs of its residents. Photograph by Louis Fox from Sustainable Revolution: Permaculture in Ecovillages, Urban Farms, and Communities Worldwide published by North Atlantic Books, copyright © 2014 by Juliana Birnbaum and Louis Fox. Reprinted by permission of North Atlantic Books.

Conclusion

A fractal is a geometric pattern, often found in nature that is repeated at ever-smaller scales to produce irregular shapes and structures. If you were to zoom in on a part of the fractal, you would see the same pattern as the whole, with slight variations, repeated to infinity.

Each of the initiatives described here depicts what permaculturists call a "node," or center of human activity. Each node can be seen as an information fractal, featuring patterns that encode a great deal of knowledge and observation. Like the fractal structures that form the blueprints of our universe, the emerging whole may be made up of a pattern that is repeated in larger and smaller scales, each slightly unique and locally adapted.

A vision of a sustainable city is being prevealed to us within the fractal network of these present-day movements. The global urban retrofit to come involves

Figure 6.7 EcoVillage at Ithaca. At the EcoVillage at Ithaca in New York, the central
corridor between homes is car-free, so that front yards open up onto communal space. This not only allows neighbors to connect more, but makes
play much safer for children. Photograph by Louis Fox from Sustainable
Revolution: Permaculture in Ecovillages, Urban Farms, and Communities
Worldwide published by North Atlantic Books, copyright © 2014 by Juliana Birnbaum and Louis Fox. Reprinted by permission of North Atlantic
Books.

changes on multiple levels, with the goal of creating circular materials economies
that are designed for zero waste.

When I look at my school-aged children, I often feel overwhelmed to think of
what their generation will face in dealing with the effects of climate change coupled with a peak in global population numbers. In the United States, the Trump
era has been a dark age in a multitude of ways, as the federal government slid
backwards into climate change denial and clings desperately to the outmoded –
and ultimately very dangerous – fossil-fuel-driven capitalist economy. Yet, this
has forced cities around the country to step up and develop their own strategies for
meeting carbon goals and responding to the climate crisis. As of present, over 500
US cities have made climate action plans or commitments (Climateactiontracker.
org 2019).

Figure 6.8 EcoVillage at Ithaca. In alignment with permaculture principles, the land at EcoVillage at Ithaca is zoned in a way that accounts for both human and natural patterns of use. About 10% is developed as clustered housing and community buildings, leaving 90% as green space in the form of deciduous forest and farms. Photograph by Louis Fox from Sustainable Revolution: Permaculture in Ecovillages, Urban Farms, and Communities Worldwide published by North Atlantic Books, copyright © 2014 by Juliana Birnbaum and Louis Fox. Reprinted by permission of North Atlantic Books.

In "The Farms of the Future," young American writer and ecologist Isabel Marlens writes about the ripple effect of vibrant local economies:

> When people and local governments reclaim their resources, it weakens the economic power of corporations, and strengthens democracy. Not only that, but societies that don't leave people desperate are societies where people are less likely to turn against each other. There is the potential for a decline in xenophobia and conflict, and for diverse communities to unite instead, against the real enemies: those who profit enormously off a system of gross injustice and inequality.
>
> (2019)

Sixteen-year-old Greta Thunberg, founder of the Sunrise Movement, has galvanized youth across borders to rally in order to pressure world governments to take

Figure 6.9 Resident at Ithaca. The permaculture principle of creatively responding to change is illustrated by this pergola at EcoVillage at Ithaca. Grape vines and foliage provide shade in summer, fruit in fall, and when leaves fall in winter allow for solar gain in the home. Photograph by Louis Fox from Sustainable Revolution: Permaculture in Ecovillages, Urban Farms, and Communities Worldwide published by North Atlantic Books, copyright © 2014 by Juliana Birnbaum and Louis Fox. Reprinted by permission of North Atlantic Books.

Figure 6.10 Children climbing, EcoVillage at Ithaca. Growing up at the EcoVillage at Ithaca gives children access to a community where play can be spontaneous and nature is easily accessible. Photograph by Louis Fox from Sustainable Revolution: Permaculture in Ecovillages, Urban Farms, and Communities Worldwide published by North Atlantic Books, copyright © 2014 by Juliana Birnbaum and Louis Fox. Reprinted by permission of North Atlantic Books.

Figure 6.11 Returning traditional agriculture to Hawaii. A permaculture project on the island of Oahu in Hawaii has restored some pre-contact kalo (taro root) pondfields. These ancient Hawaiian irrigation systems, once common on the mountainsides, respectfully borrowed water from a primary mountain stream, to flow down through a cascading pattern of kalo ponds that led the water ultimately back to its original stream where it could finish its journey to the sea. Kalo was a staple vegetable in the traditional Hawaiian diet. Photograph printed with permission by Louis Fox.

action. In the space of a little over a year, she went from sitting alone with her sign outside government buildings in her home country of Sweden to being recognized as the leader of a worldwide climate strike involving 163 countries on September 20, 2019. My children and I were among the estimated five million to seven million voices calling for climate action that day. Greta sailed from Sweden to address the UN Climate Action Summit in New York, closing by her fiery appeal to government leaders to take action with:

> You are failing us. But the young people are starting to understand your betrayal. The eyes of all future generations are upon you. And if you choose to fail us I say we will never forgive you. We will not let you get away with this. Right here, right now is where we draw the line. The world is waking up. And change is coming, whether you like it or not.

(2019)

The post-capitalist city embodies the waking up and the transformation that Greta invokes here, and it is a vision that can only come through a powerful synergy of local and global action. It is, as Naomi Klein puts it, the time for "astronaut's eye-view environmentalism" to pass, with:

> a new movement rising to take its place, one deeply rooted in specific geographies but networked globally as never before. Having witnessed the recent spate of big failures, this generation of activists is unwilling to gamble with the precious and irreplaceable.
>
> (2015:290)

And in their determination, I can already see them – Greta's generation, my children among them – building the city we see emerging on these pages, the ark to carry them through the turbulent seas of our global crisis.

References

Birnbaum, J. & Fox, L., 2014, *Sustainable Revolution: Permaculture in Ecovillages, Urban Farms and Communities Worldwide*, North Atlantic Books, Berkeley.

Boer, B.J. & Kaan, H.F., 2006, *Passive houses: Achievable concepts for low CO_2 housing*, ECN Publications, Petten, Holland. 9RX99069019.

Climateactiontracker.org, 2019, *USA Climate Action Tracker*, [online], https://climateactiontracker.org/countries/usa/.

Cohen, S., 2018, *The sustainable city*, Columbia University Press, New York, NY.

The Conversation, 2019, *Urban Agriculture: What U.S. Cities Can Learn From Cuba*, [online] US News & World Report, viewed 9 January 2020, fromwww.usnews.com/news/cities/articles/2019-02-13/urban-agriculture-what-us-cities-can-learn-from-cuba.

Dregger, L., 2015, *Ecovillage Summit in Dakar, Senegal – Global Ecovillage Network*, [online] Global Ecovillage Network, viewed 9 January 2020, fromhttps://ecovillage.org/ecovillage-summit-in-dakar-senegal/.

Ecovillageithaca.org, 2020, *Ecovillage at Ithaca | Website for Ecovillage at Ithaca and its educational non-profit Learn@EcovillageIthaca*, viewed 9 January 2020, fromhttps://ecovillageithaca.org/.

Evans, M., 2019, 'HUD's Ben Carson: More should follow Minneapolis and phase out single-family zoning', *Star Tribune*, viewed 9 January 2020, fromwww.startribune.com/u-s-housing-secretary-ben-carson-visits-minnesota-for-first-time-as-federal-official/511485922/.

Foderaro, L., 2012, 'In Rooftop Farming, New York City Emerges as a Leader', *The New York Times*, [online] 12 July, viewed 9 January 2020, fromwww.nytimes.com/2012/07/12/nyregion/in-rooftop-farming-new-york-city-emerges-as-a-leader.html.

Global Ecovillage Network, 2016, *Ecovillages – Global ecovillage network*, viewed 9 January 2020, fromhttps://ecovillage.org/projects/.

Hawken, P., 2017, *Drawdown: The most comprehensive plan ever proposed to reverse global warming*, Penguin Books, New York, NY.

Hiller, M., 2008, 'Interview with Juliana Birnbaum', *Sustainable [R]evolution*, 11 June.

Houston, W., 2019, 'Outage-prone West Marin mulls microgrids to create an Island of power', *Marin Independent Journal*, viewed 9 January 2020, from www.marinij.com/2019/05/29/microgrids-in-west-marin/.

Johnson, C., 2014, 'District heating: A hot idea whose time has come', *The Guardian*, viewed 9 January 2020, fromwww.theguardian.com/cities/2014/nov/18/district-heating-a-hot-idea-whose-time-has-come.

Klein, N., 4 August 2015, *This changes everything: Capitalism vs. the climate*, p. 133, Reprint edn., Simon & Schuster, New York, NY.

Marlens, I., 2019, *The farms of the future*, [online] Common Dreams, viewed 9 January 2020, fromwww.commondreams.org/views/2019/06/05/farms-future.

Nationalgeographic.com, 2017, *This city aims to be the world's greenest*, viewed 9 January 2020, fromwww.nationalgeographic.com/environment/urban-expeditions/green-buildings/green-urban-landscape-cities-Singapore.

Permaculture magazine, 2013, *Permaculture Magazine*, viewed 9 January 2020, from www.permaculture.co.uk/articles/womens-ecovillage-colombia-nashira.

REHVA, 2018, *REHVA journal 01/2018 – India's first district cooling system at GIFT city*, [online], www.rehva.eu/rehva-journal/chapter/indias-first-district-cooling-system-at-gift-city.

Rosan, C. & Perasall, H., 2017, *Growing a sustainable city? The question of urban agriculture*, pp. 6–7, University of Toronto Press, Toronto, Canada.

Roudman, S., 2013, 'The Green-Building Racket', *The New Republic*, viewed 10 January 2020, from https://newrepublic.com/article/113942/bank-america-tower-and-leed-ratings-racket.

Roy, A., 2010, 'The trickledown revolution', *Outlook India Magazine*, viewed 9 January 2020, from www.outlookindia.com/magazine/story/the-trickledown-revolution/267040.

Sustainability Times, 2019, *Can ecovillages be an answer to environmental challenges?*, viewed 9 January 2020, fromwww.sustainability-times.com/clean-cities/can-ecovillages-be-an-answer-to-future-environmental-challenges/.

Taggart, J., 2009, 'Inside an ecovillage: Born of aligned ecological values and design, ecovillages are found in over 70 countries around the world', *Alternatives Journal* 35(5), 20+. *Gale Academic Onefile*, viewed 9 January 2020.

There's Community and Consensus: But It's No Commune, 2018, *The New York Times*, [online] 20 January, viewed 9 January 2020, from www.nytimes.com/2018/01/20/business/cohousing-communities.html.

Thunberg, G., 2019, 'If world leaders choose to fail us, my generation will never forgive them | Greta Thunberg', [online] *The Guardian*, www.theguardian.com/commentisfree/2019/sep/23/world-leaders-generation-climate-breakdown-greta-thunberg.

Tsui, B., 2019, 'Hong Kong's skyline farmers', [online] *The New Yorker*, www.newyorker.com/tech/annals-of-technology/hong-kongs-skyline-farmers.

unenvironment.org, 2019, [online], www.unenvironment.org/news-and-stories/story/skys-limit-architects-design-un17-eco-village-copenhagen.

United Nations, Revision of World Urbanization Prospects, 2018, World Urbanization Prospects, [online] United Nations, p. 1, viewed 6 July 2019, from www.un.org/development/desa/publications/2018-revision-of-world-urbanization-prospects.html.

USA TODAY, 2012, *In U.S. building industry, is it too easy to be green?*, viewed 10 January 2020, from www.usatoday.com/story/news/nation/2012/10/24/green-building-leed-certification/1650517/.

Usda.gov, 2019, *Vertical Farming for the Future*, [online], www.usda.gov/media/blog/2018/08/14/vertical-farming-future.

Walker, L., 2005, *EcoVillage at Ithaca: Pioneering a sustainable culture*, New Society Publishers, Gabriola, BC.

7 Urban commons

Toward a better understanding of the potentials and pitfalls of self-organized projects

Mary H. Dellenbaugh-Losse

Introduction: necessity and the re-invention of the urban commons

Suddenly, it feels like the topic of urban commons is everywhere. To be true, the topic of urban commons has exploded over the last decade and a half, leading to a wealth of academic and activist-led scholarship on the topic (i.e. Le Goix & Webster 2006; Lee & Webster 2006; Helfrich & Heinrich-Böll-Stiftung 2009, 2012; Hess 2009; Foster 2009, 2013; Parker & Johansson 2011; Bollier, Helfrich & Heinrich-Böll-Stiftung 2012; Exner & Kratzwald 2012; Newman 2013; Borch & Kornberger 2015; Dellenbaugh *et al.* 2015; Stavrides 2016; Chatterton 2016; Parker & Schmidt 2017; Shareable 2017; Bollier & Helfrich 2019; Dellenbaugh-Losse, Zimmermann & de Vries 2019). What are urban commons and what is it about the last 15 years that has given rise to the current debates about commons in cities?

Urban commons can be defined as "resources in the city which are managed by the users in a non-profit-oriented and prosocial way" (Dellenbaugh-Losse *et al.* 2019). Despite a broad conceptualization of commons in the academic literature, they generally share two main characteristics. Firstly, urban commons projects prize the use value of resources, i.e. the practical, everyday value of the resource for its users, above their exchange value, i.e. the price one could receive for selling the resource or use rights on the free market. Commons projects focus on long-term, sustainable, user-centric solutions in which the value created by a project is not extracted but rather fed back into the process of solidification and expansion. Secondly, urban commons projects reverse the logic of the capitalist market and the state, both of which force the citizen into the subject/consumer role. Instead, they empower citizens to address their own needs and co-produce solutions to urban issues. In the process, the "commoners" gain valuable skills in advocacy and civic participation.

In the context of this collected work, I see two main turning points which I would point to as "moments of necessity" from which the "re-invention" of urban commons was born. In 2007–2008, the global financial crisis sent shockwaves through the global(ized) economy, resulting in widespread austerity measures and the reduction and elimination of public goods and services from Athens to Manchester. In some ways, one could regard the effects of the global financial crisis as merely an extreme and sudden concentration of the ills of neoliberal fiscal policy which had been gradually accruing for decades. Urban commons, some explicitly described as such and some not, emerged from this situation as both emancipatory and "self-help" measures. Commoners tended neglected public parks, offered community health care, and developed new housing alternatives for victims of foreclosure (Dellenbaugh *et al.* 2015). But what has happened to these emancipatory projects and their promise of authentic resident participation? As the market and the state have recovered from their collective tailspin, these projects have often become the victims of privatization or other forms of enclosure. This change is fundamentally bound up with what I would describe as the second "moment of necessity."

Cities, especially in North America and Europe, are increasingly experiencing a crisis of affordability (e.g. following Florida 2017). Over the last three decades, entrepreneurial city tactics have led to the privatization of municipal housing, public utilities, and public services through sales or outsourcing, and massive reductions in municipal landownership. Under conditions of increasing existential uncertainty in urban centers, urban commons have become not only a way to secure (affordable) access to resources, but also a rallying cry against the waves of privatization of once-public resources that are sweeping above all growing cities. It is into this niche that cries of the "right to the city" (Lefebvre 1996) have been shouted from London to San Francisco. The increasing financialization of urban land and urban resources has made the diverse roles that the market plays in the city today clearer than ever: the market as a spatial sorting mechanism for more- and less-well-off urban residents, urban real estate's increased role as lines in investment portfolios, touristification, gentrification, the next big-event-driven infrastructure project . . . It's not hard to understand the urban populace's desire to wrest the control over at least some aspects of their day-to-day lives and environment back from the state and market.

Urban commons, therefore, can be described as having had their genesis in *both* the absence of the state and market ("citizen self-help") *and* in the overbearing presence of the state and market (resistance to enclosures resulting from commodification, financialization, or new legal circumstances), both of which have directly resulted from the neoliberal policies of the 1970s and 1980s. Perhaps the question should therefore be: why did it take so long for the topic of urban commons to emerge? I would argue that it required the shock and collective pain of the global financial crisis to bring the accrued effects of decades of austerity and privatization into sharp focus. The result is the wave of scholarship we see today.

Urban commons have the potential to be more inclusive and guarantee accessibility to resources for a wider group of people in the city than is the case for either public or consumer goods, but it doesn't have to be that way. Building on the definition stated at the outset of this chapter, I would now like to examine both the benefits and risks associated with the economic and social/civil aspects of urban commons projects.

Benefit: use over profit

The concept of urban commons contains a fundamental shift in the understanding of resources away from commodification and financialization. As I recently examined elsewhere (Dellenbaugh-Losse 2019a, 2019b), the "transaction" behind urban commons demonstrates two main differences from traditional market transactions.

First, we can understand urban commons as a form of resource use and co-management in which the user contributes time and effort and (therefore) pays less for the resource use (and gets a social added benefit as well). This stands in contrast to traditional capitalist models, in which the worker converts her labor into money which she then uses to buy products in the market. So, while in capitalist models, one converts human capital into financial capital, which can then be exchanged for goods and services, in commons, commoners invest their human, social, and financial capitals *directly* and both receive goods and services and increase their social capital (Dellenbaugh-Losse 2019a). Commons thus omit part of the capitalist process; instead of the added step of converting one form of capital into another (see also Bourdieu 1986), commons foresee the direct investment of time and human and social capital into the project. This "elimination of the middleman" also removes an opportunity for profit extraction, the second main difference between commons and capitalist transactions.

Following standard capitalist business models, the price of the widgets that our hypothetical worker can buy on the open market is comprised not only of the material, labor, and overhead costs inherent to producing that widget but also some portion of profit, which is then available to be extracted from the system, for example in the form of stock dividends or manager bonuses. In contrast, commons projects concentrate on the use value of resources, i.e. *only* the material, labor, and overhead costs required to produce and/or maintain the resource in question. By investing in long-term, sustainable, closed systems (at least with regard to the various capital investments), commoners shift the focus from concentrated financial benefits for a select few to shared diverse benefits for all involved.

Risk: enclosure and responsibilization

Both the state and the market recognize these added benefits and have been known to attempt to capture these positive externalities for themselves. As soon as added value is created, the resource becomes (more) attractive for market and state actors – the case we saw above with the enclosure of "citizen self-help" commons during the recovery from the global financial crisis. The risks inherent to commons' "unrealized gains" (Tuovila 2019) and proactive citizen engagement fall into two categories: enclosure and responsibilization.

The term "enclosure of the commons" is derived from the enclosure movements in northern Europe (most notably in England) more than five centuries ago (Linebaugh 2009). In this process, commonly held lands, which were an important source of food and fuel for the peasants, were fenced in and commodified.

Today, "enclosure" refers to the partial or total loss of accessibility to a resource. In the city, enclosure can take the form of privatization or commercialization or result from new laws (Lee & Webster 2006; Hodkinson 2012).

In addition to the enclosure of excludable resources, i.e. resources to which access can be denied, technological advances and new laws may pave the way for new forms of enclosure, much like the invention of the land registry paved the way for the original enclosure of the commons (Harford 2017). Law changes which put new restrictions on existing projects to limit their opening times or barriers to access which allow the monetization of content (such as DRM or pay-walls) are examples of emergent or new enclosures of commons. As opposed to a simple restriction of access or partout privatization of a resource, in the city, the enclosure of the urban commons refers to a much wider range of practices which restrict access, from blatant commercialization to the introduction of police-enforced dusk-to-dawn curfews.

In addition to enclosure practices by both the state and the market, commons are also at risk of "responsibilization" by the state, practices in which urban com-mons projects are instrumentalized to fulfill obligations which were once the pur-view of the state or municipality. The term responsibilization refers to the transfer of responsibility for a problem or issue from the state agency or authority to the individual or community (following Wakefield & Fleming 2009); responsibiliza-tion is deeply connected to neoliberal and entrepreneurial governance strategies (Ilcan & Basok 2004; Goddard 2012; Pyysiäinen, Halpin & Guilfoyle 2017).

In the case of urban commons, "citizen self-help" can mean the assumption of tasks, goods, and services which were once the responsibility of the state and its administration. However, in contrast to citizen empowerment (following Arnstein 1969), responsibilization merely outsources (or crowdsources) formerly central-ized processes, such as crime prevention, *without* an increase in political or civic agency on the part of the citizen. Thus, as opposed to an increase in citizen control *and* citizen responsibility, responsibilization confers only one half of the equation, retaining the control with the state.

Benefit: from consumers to co-producers

From a social and civic perspective, commons fundamentally change the role of the citizen or resident from a consumer of goods and services provided by the state and/or market to a co-producer of co-used goods and services. In the con-text of participative democracy (following, e.g., Arnstein 1969), this constitutes a shift in the understanding of citizens' roles from "citizen-consumers" to citizen co-producers. To understand exactly what is meant here, we need to take a quick step back.

The concept of the "citizen-consumer" and the conflation of economic liberal-ism and political freedom in the western world found their genesis during the Cold War. The bundling of the political form, democracy, and the economic form, free market capitalism, paralleled the inherent combination of the communist state and its state-run economy, and pitted these as polar opposites in the war for ideological dominance (Dellenbaugh-Losse 2020). The biggest champion of this idea was Milton Friedman, who received the Nobel Memorial Prize in Economics in 1976. Through his decades-long work, he effectively propounded the theory that economic and political freedoms were inherently linked; the only economic system which could guarantee individual political freedoms, he argued, was free market capitalism (Friedman 2002). His work was highly influential in the development of neoliberal policies during the Reagan and Thatcher eras and was an important ideological springboard for western Cold War dogma.

Furthermore, in western Cold War propaganda, "political action and freedom of electoral choice were often conflated with consumer choice. The amount and variety of products on stores shelves was cited as evidence of the democracy and superiority of the western 'way of life'" (Pugh 2014:209). Again, political agency was set equal with economic agency under the heading of "choice is choice, and more choice is always better." Cold War propaganda and Friedman's ideas (which were picked up and promoted by official channels) served to naturalize the concept in the West that individual freedom and political agency were inherently linked to consumerism, and especially to free market capitalism. This ideological foundation promoted the consumer-driven individualism seen in the 1970s and 1980s (following Wolfe 1976). The concentration on freedom, and particularly individual freedom, also served as a logical justification for the deregulation that took place in North America and the United Kingdom in the 1980s and 1990s.

The role of the consumer is, however, inherently passive. The market creates products and marketing bolsters demand for them; the role of the consumer is merely to select from the choices available to her. Seen through the lens of Arnstein's ladder of participation (1969), the role of the consumer is one of non-participation or, at most, tokenism. Free market capitalism assumes that the "invisible hand" will satisfy the needs of the masses; however, tasking the market with satisfying the public's needs not only disadvantages less profitable goods and services but also inherently places the consumer in the subject role. Commons invert this logic, instead placing the citizen in the role of the co-producer and the collective co-production and co-maintenance of the resource center stage. Through the highly social and empowering process of commoning, the commoners transition from subjects to agents of change in which collectivization actually *increases* political agency. Indeed, the deliberation and negotiation inherent to commoning and the wide range of soft skills that commoners learn along the way form a solid foundation for other forms of civil engagement and participation (following Dellenbaugh-Losse *et al.* 2019).

Commons represent therefore not only a movement against crises of affordability and for equitable resource access in cities, but also a new social drive away from the atomized individualism of the 1980s, which was in part brought about by predatory markets emboldened and empowered by neoliberal deregulation. American and western European societies have largely shifted away from traditional and historical social and political structures (following Putnam 2000). However, alternative structures have cropped up in their place (following Lemann 1996); commons projects are one example of these. Instead of accepting existing, pre-determined structures, today's proactive urban residents are creating their own. This represents a confluence of "citizen self-help" and affordability reasoning; proactive citizens create new forms of access to resources and thus ensure their continued access to said resource. They both create "social regime[s] for managing shared resources" and "forg[e] a community of shared values and purpose" (Clippinger and Bollier, quoted in Hess 2009:35). It is precisely this "community of shared values and purpose" that reflects not the loss in social capital that Putnam described, but rather a transference of these practices to new structures.

These tendencies reflect a wider change in today's cities which has been brought about by deregulation, free trade, free movement, and globalization. The increase in mobility inherent to globalization is acutely apparent in today's cities; many of those affected by civil policies do not have the right(s) to change them. Urban commons projects can provide the opportunity for those otherwise disenfranchised from structural political change to co-produce urban interventions on a smaller, less formal scale. Urban commons projects have the potential to create even footing between citizens and non-citizens, and allow those affected by policies to have a voice in how they are structured.

"Citizen self-help," perhaps better termed "resident self-help," is not only the result of the effects of the global financial crisis. It is also a user-led solution to the rollback of the welfare state and state provision of goods and services. The transition from the welfare state to privatization to commons represents a set of shifts in access to and control over resources. In a first step, neoliberal and entrepreneurial tactics privatized welfare goods and services which had been only available to citizens and those fulfilling certain legal criteria. These goods and services, now "freed" from state control, were now subject to different controls – no longer to do with passports, but rather bankbooks. Commons remove these resources from both state and market forces and their various barriers to entry; however, the self-selection and self-regulation of member and user groups is not without its own risks and challenges.

Risk: co-option and enclavism

The question is: who is invited to join in this process? Several of the case studies for *The Urban Commons Cookbook* stated that they explicitly sought like-minded people with similar values (Dellenbaugh-Losse *et al*. 2019). Joining forces with others with similar worldviews seems like a logical step; for the projects interviewed, it removed a layer of complexity from the commoning process. However, taken to the extreme, this tendency could lead to co-option of urban resources by certain groups and enclavism. To talk about these risks in an informed way, we again need to take a step back.

Many urban commons are excludable resources; access can be denied through fences, membership, passwords, or any number of other means; this is especially true of landed commons. In these cases, the commoners represent a self-selected group who co-determine who gets to join and who does not. Seen from this perspective, commoning has the potential to be a lot less inclusive than it is usually touted to be. Indeed, in the case of the "Right to the City" movement in Hamburg, one study found that more than 90% of the participants were white, educated, middle-class German natives (Kortus 2014).

It's well known that "birds of a feather flock together." This concept is formally known as the homophily principle; simply put, "similarity breeds connection" (McPherson, Smith-Lovin & Cook 2001:415). In group dynamics, however, large numbers of members of one type or category can have an implicit exclusionary effect for people from other categories. Taken one step further, group members can also consciously or unconsciously act as "gatekeepers," successfully reducing the homogeneity of the group. Susanne Frank's research about enclavistic tendencies among the well-educated white middle class in Germany, what she describes as "inner suburbanization" in cities (Frank 2012, 2014), is instructive. In her case studies, well-educated middle class families used discourses of inclusion (e.g. "Traveplatz for everyone"; Frank 2012, p. 74) to restructure the neighborhoods they had moved into, functionally excluding other groups; in this case, "Traveplatz for everyone" actually sought to redesign the square in question to make it more child-friendly and less welcoming for "undesirable" groups such as punks, public drinkers, the homeless, and teenagers. A recent study in Paris examined similar tendencies in the urban commons (Newman 2013).

Without going too far down the urban sociology rabbit hole, it is important to consider not only the tactics but also the relative power of different groups. The academic and activist roots of the urban commons naturally lead to a larger number of socially privileged groups involved in commons projects. These groups, in particular the white middle class, have the most to lose from the current tendencies of socioeconomic polarization. The crisis of affordability affects them the most acutely; suddenly their relative prosperity seems not to go as far as it used to.[1] The white middle class is, however, already empowered and has high political, social, and human capital which it can bring to bear in and through community and neighborhood projects. Their shared struggle and experience and the process of commoning promote the creation of bonding social capital (Putnam 2000, p. 22),[2] while homophily can act to limit the heterogeneity of new members (McPherson *et al.* 2001). The result is homogenous groups of commoners whose group composition is maintained by relatively effective semi-porous social membranes between the community and the commons project.

Thus, if we consider that one of the main implications behind the right to the city is the access to resources, especially those required for a decent existence, then we would be justified in critically examining who is claiming the right to the city and who is not (able to). Fear of loss leads to the hardening of group boundaries (Gest 2016; Mutz 2018). Similarly, perceived scarcity can lead to the conscious or unconscious co-option of the resource for in-group use by those with higher amounts of social, human, and/or political capital. Under these conditions, urban commons projects could in fact contribute to, not ameliorate, resource scarcity and access equity problems in cities.

Avoiding common(s) pitfalls

Commoning is a discursive process through which co-management is negotiated and important social bonds and controls are forged. As the case studies for *The Urban Commons Cookbook* showed, it is also an iterative process, constantly evolving relative to external forces and internal changes. For this reason, it is thinkable that a large range of urban commons projects will find themselves faced, possibly even repeatedly, with the challenges described earlier over the course of their projects. The inherent agility of the commoning process, however, allows the commoners to recognize and react, shifting the course of the project back towards one which is more stable and inclusive. A range of checks can serve as helpful tools to analyze whether commons projects are subject to the risks discussed in this chapter.[3]

The first check involves **access**. Commoners should keep abreast of commodification trends and changes to regulations which might affect access to their project (e.g. new restrictions on serving food and/or alcohol or opening times) and periodically check whether the group is truly open for all. A simple inclusion/participation/representation check can suffice for the latter (following Fahrun, Zimmermann & Skowron 2016:62). Commoners should ask themselves: what barriers exist to joining our project? What opportunities do people have to participate? How can we make sure that a diverse group of voices is being heard? If there is internal resistance to openness, what is it based on?

The second check involves a critical examination of the **balance between responsibility and agency**. If the project fulfills an important function, do those with the responsibility also have the ability to change the rules that affect them? What degree of involvement to commoners have with the politicians and decision-makers whose decisions affect them (following Arnstein 1969)? How can participative structures be reworked into mutually beneficial trans-sectoral cooperations?

Finally, the third check requires commoners to carefully examine their **goals and motivation**. Commoners are recommended to examine both why they are undertaking their project and also how they are transferring these values and purposes into action. Commoners should not only critically and honestly examine why they are in the project – what it is and what it stands for – but also the ways in which the implementation of the project may in fact be at odds with the mission, goals, values, and beliefs at its core.

Conclusion

Urban commons will likely continue to grow in importance, both in academic and activist circles. It is the author's opinion that practice-based research presented in non-academic language is highly needed to promote knowledge transfer across silos and between the academic and activist community. After all, urban commons are not just a "hot" academic topic; for some, they represent the difference between existential precariousness and a decent life.

Notes

1 This is particularly true for young urban professional parents (*Yupps*) and double-income households with kids (*Diwiks*) (following Frank 2012).
2 In *Bowling Alone*, Robert Putnam examines two forms of social capital, which he describes as "bonding" and "bridging." Bonding social capital increases the solidarity and cooperation *within* groups, while bridging social capital increases the connections *across and between* groups (Putnam 2000).
3 These and other tools are included in the 2020 book, *The Urban Commons Cookbook: Strategies and Insights for Creating and Maintaining Urban Commons*.

References

Arnstein, S.R., 1969, 'A ladder of citizen participation', *Journal of the American Planning Association* 35(4), 216–224.
Bollier, D. & Helfrich, S., 2019, *Free, fair, and alive: The insurgent power of the commons*, New Society Publishers, Gabriola Island.
Bollier, D., Helfrich, S. & Heinrich-Böll-Stiftung (eds.), 2012, *The wealth of the commons: A world beyond market and state* Levellers Press, Amhearst.
Borch, C. & Kornberger, M. (eds.), 2015, *Urban Commons: Rethinking the City* Routledge, New York, NY.
Bourdieu, P., 1986, 'The forms of capital', in J. Richardson (ed.), *Handbook of Theory and Research for the Sociology of Education*, Greenwood, New York, NY, pp. 241–258.
Chatterton, P., 2016, 'Building transitions to post-capitalist urban commons', *Transactions of the Institute of British Geographers* 41(4), 403–415.
Dellenbaugh, M. et al., (eds.), 2015, *Urban Commons: Moving Beyond State and Market*, Birkhäuser Verlag, Basel.
Dellenbaugh-Losse, M., 2020, *Inventing Berlin. Architecture, Politics and Cultural Memory in the New/Old German Capital Post-1989*, Springer International Publishing, Cham.
Dellenbaugh-Losse, M., 2019a, *Urban commons as part of modern urban capitalism? A thought experiment*, viewed 17 July 2019, fromhttps://urban-policy.com/2019/06/03/urban-commons-as-part-of-modern-urban-capitalism-a-thought-experiment/.
Dellenbaugh-Losse, M., 2019b, *What unites urban commons and the affordable housing crisis? The problem of profit extraction*, viewed 17 July 2019, from https://urban-policy.com/2019/06/25/the-problem-of-profit-extraction/.
Dellenbaugh-Losse, M., Zimmermann, N.-E. & de Vries, N., 2019, *The Urban Commons Cookbook: Strategies and Insights for Creating and Maintaining Urban Commons*, Berlin.
Exner, A. & Kratzwald, B., 2012, *Solidarische Ökonomie & Commons*, Mandelbaum, Vienna.
Fahrun, H., Zimmermann, N.-E. & Skowron, E., 2016, *The Initiative Cookbook – Homemade Civic Engagement: An Introduction to Project Management*, www.mitost.org/editions/initiativecookbook/THK/Cookbook_English/THK-Cookbook_English.pdf.
Florida, R.L., 2017, *The new urban crisis: How our cities are increasing inequality, deepening segregation, and failing the middle class – and what we can do about it*, Basic Books, New York, NY.
Foster, S.R., 2009, 'Urban informality as a commons dilemma', *Inter-American Law Review* 40(2), 301–325.

Foster, S.R., 2013, 'Collective action and the urban commons', *Notre Dame Law Review* 87(1), 57–134, https://scholarship.law.nd.edu/ndlr/vol87/iss1/2.

Frank, S., 2012, 'Reurbanisierung als innere Suburbanisierung', in A. Hill, A. Prossek & A. Lindner (eds.), *Metropolis und Region. Herausforderungen für Stadtforschung und Raumplanung zu Beginn des 21. Jahrhunderts*, pp. 69–80, Rohn Verlag, Detmold.

Frank, S., 2014, 'Innere Suburbanisierung als Coping-Strategie: Die „neuen Mittelschichten"in der Stadt', in P.A. Berger *et al.* (eds.), *Urbane Ungleichheiten*, pp. 157–172, Springer, Wiesbaden.

Friedman, M., 2002, *Capitalism and Freedom*, 40th Anniv., University Of Chicago Press, Chicago.

Gest, J., 2016, *The New Minority: White Working Class Politics in an Age of Immigration and Inequality*, Oxford University Press, New York, NY.

Goddard, T., 2012, 'Post-welfarist risk managers? Risk, crime prevention and the responsibilization of community-based organizations', *Theoretical Criminology* 16(3), 347–363.

Harford, T., 2017, *Fifty things that made the modern economy*, Little, Brown Book Group, London.

Helfrich, S. & Heinrich-Böll-Stiftung (eds.), 2009, *Genes, Bytes and Emissions: To Whom Does the World Belong?* Berlin, www.boell.org/web/148-576.html.

Helfrich, S. & Heinrich-Böll-Stiftung (eds.), 2012, *Commons: Für eine neue Politik jenseits von Markt und Staat*, Transcript, Bielefeld.

Hess, C., 2009, 'Mapping the New Commons', in *12th Biennial Conference of the International Association for the Study of the Commons*, http://ssrn.com/abstract=1356835

Hodkinson, S., 2012, 'The new urban enclosures', *City* 16(5), 500–518.

Ilcan, S. & Basok, T., 2004, 'Community government: Voluntary agencies, social justice, and the responsibilization of citizens', *Citizenship Studies* 8(2), 129–144.

Kortus, M.S., 2014, *Recht auf Stadt – Zwischen Lefebvres Vision und der Praxis städtischer sozialer Bewegungen im 21. Jahrhundert. Das Beispiel das Recht auf Stadt Netzwerkes Hamburg*, Humboldt-Universität zu Berlin.

Lee, S. & Webster, C., 2006, 'Enclosure of the urban commons', *GeoJournal* 66(1–2), 27–42.

Lefebvre, H., 1996, *Writings on Cities*, Blackwell Publishing, Oxford.

Le Goix, R. & Webster, C., 2006, 'Gated communities, sustainable cities and a tragedy of the urban commons', *Critical Planning* 13, 41–64.

Lemann, N., 1996, 'Kicking in groups', *The Atlantic*, April, www.theatlantic.com/magazine/archive/1996/04/kicking-in-groups/376562/

Linebaugh, P., 2009, *The Magna Carta manifesto: Liberties and commons for all* University of California Press, Berkeley.

McPherson, M., Smith-Lovin, L. & Cook, J.M., 2001, 'Birds of a Feather: Homophily in Social Networks', *Annual Review of Sociology* 27, 415–444, www.jstor.org/stable/2678628

Mutz, D.C., 2018, 'Status threat, not economic hardship, explains the 2016 presidential vote.', *Proceedings of the National Academy of Sciences of the United States of America* 115(19), E4330–E4339.

Newman, A., 2013, 'Gatekeepers of the urban commons? Vigilant citizenship and neoliberal space in multiethnic paris', *Antipode* 45(4), 947–964.

Parker, P. & Johansson, M., 2011, *The uses and abuses of Elinor Ostrom's concept of commons in urban theorizing*, Presented at the International Conference of the European Urban Research Association (EURA) 2011 - Cities without Limits (23–25 June 2011 in Copenhagen).

Parker, P. & Schmidt, S., 2017, 'Enabling urban commons', *CoDesign* 13(3), 202–213.

Pugh, E., 2014, *Architecture, Politics, & Identity in Divided Berlin*, University of Pittsburgh Press, Pittsburgh.

Putnam, R.D., 2000, *Bowling Alone: The Collapse and Revival of American Community*, Simon & Schuster, Inc., New York, NY.

Pyysiäinen, J., Halpin, D. & Guilfoyle, A., 2017, 'Neoliberal governance and "responsibilization" of agents: Reassessing the mechanisms of responsibility-shift in neoliberal discursive environments', *Distinktion: Journal of Social Theory* 18(2), 215–235.

Shareable, 2017, *Sharing Cities: Activating the Urban Commons*, www.shareable.net/sharing-cities/

Stavrides, S., 2016, *Common space: The city as commons*, Zed Books Ltd., London.

Tuovila, A., 2019, *Definition: Unrealized Gain*, *Investopedia*, viewed 1 August 2019, from www.investopedia.com/terms/u/unrealizedgain.asp.

Wakefield, A. & Fleming, J. (eds.), 2009, 'Responsibilization', in *The SAGE Dictionary of Policing*, SAGE Publications Ltd., London.

Wolfe, T., 1976, 'The "Me" Decade and the Third Great Awakening', *New York Magazine*, August.

8 Counteracting the negative effects of real-estate-driven urbanism + empowering the self-constructed city

David Gouverneur

Introduction

This article intends to share some ideas helping counteract the negative aspects of the capital-driven urbanism, addressing social inclusion and environmental soundness, with a hybrid and transitional approach that merges planning and design with the dynamism of the informal city. My take on the post-capitalist city derives from a broad knowledge of Latin-American urbanism, particularly on the processes of informal urbanization, and the possibilities of guiding the emergence of new self-constructed habitats while helping them to become a dominant component of sustainable, environmentally responsible, and equitable urban scenarios.

The term post-capitalism has usually been equated to forms of Marxism, socialism, or anarchism but it may have other contemporary interpretations. It can refer to modes of production and consumption, equitable distribution of wealth or benefits (including changing values concerning both), or to modes of governance-participation, and nonconventional territorial-spatial and performative conditions.

In light of the current environmental crisis, the term may also refer to envisioning how a growing population can continue living on the planet, easing and even reversing the damage that we have already caused. Some may also associate post-capitalism to a stage in which the accumulation of wealth and goods is replaced by one of knowledge in an area of unlimited technologically based information.

Whichever the interpretation, I am particularly skeptical of "solutions" that operate on the premise that the structure that supports the existing social exclusionary and degraded environmental conditions have to be entirely replaced or destroyed before any real progress can be made. This approach is usually traumatic, leaves little room for interactions or sustainable transitions. I rather favor processes of gradual transformation operating on existing layers of histories and ecologies making the systems more resilient, stronger, diverse, and socially just (see Figures 8.1 and 8.2).

A failed "socialist revolution"

I had the opportunity to experience closely the economic, urban, environmental, and ethical meltdown of my homeland, Venezuela, during the past two decades

Figure 8.1 Informal settlement on the floodplain of the Catuche Ravine, Caracas, Venezuela.

Photo: David Gouverneur

under the leadership of a so-called Bolivarian Socialist Revolution of the Twenty-First Century. This was once a rich state oil-driven nation that received millions of immigrants from war-torn Europe, the Middle East, and Latin American countries fleeing poverty, extreme violence, and military oppression.

Figure 8.2 Informal settlement on steep terrain in Petare, Caracas, Venezuela.

Photo: Nicola Rocco, Caracas Cenital

For many – and from afar – this revolutionary experiment democratically elected in 1998 with ample support was considered a progressive socialist movement that sought to eliminate the evils of the capitalist system. It intended to eradicate corruption, reduce inequalities, and provide visibility to a segment of the country that remained marginalized from economic benefits and political decisions. This impoverished segment at the time represented 20–25% of the population, part of which were descendants of migrants from rural areas and other countries attracted to Venezuela by the jobs, services and, at the time, political stability.

After being elected, the government organized socially oriented programs, e.g., micro-clinics in the most challenged informal neighborhoods with the support of Cuban doctors; reading and writing courses for the elderly of rural origin; sports facilities; etc. Cooperative organizations were also created that would receive federal funds and provide cash to their members, surpassing the average minimum wages, provided they registered in the socialist party.

Such programs initially helped to spread the wealth of a country 90% dependent on oil revenues to the less affluent groups. But in time, workers gradually ceased performing traditional jobs like fishing, farming, artisanal skills, etc., which were productive activities that were free from the influence of large capitals corporations. Some cases were reported where the cooperatives received the financial support and invested it in dollars offshore to protect it from rampant devaluation and inflation within the country.

While democratically elected with ample support, the government gradually embraced populism, with all the foes of authoritarian regimes, resorting to media manipulation, electoral fraud, intimidation, and terror to stay in power. It also maneuvered to take control of all branches of government and the military. In the long run, these moves would disable the democratic system that had allowed it to emerge. Another aspect of the process was depriving the population of education and skills, as ignorance and handouts kept the souls tame until total political control was achieved.

Nationalizing private enterprises without compensation and making it difficult for the private sector to operate led to the closing of industries and the loss of jobs. Dismantling all income-generating initiatives, the mismanagement of the state-owned oil industry, and the dismissal of qualified public servants resulted in the economic break down of the country. As a result, there is a lack of basic services, food, medicine, electricity, potable water, as well as the surge of rampant crime and the fear of a bleak future for the majority of the population. Meanwhile, the leaders of the "Revolution" and their close supporters became filthy rich, entrenched in power, immersed in illegal activities, insensitive to the people's needs, and puppets of international both official and obscure forces.

Finally, rampant corruption, the handout of financial aid to Cuba and other political allies (in cash and oil), coupled with the collapse of oil prices, left the central government with empty pockets to continue these politically motivated policies. The lack of public funds paved the road for governmental involvement in currency control and fake monetary conversion, drug-trade, illegal mining/ tropical deforestation, and even corrupt scams for the import of food as internal production collapsed.[1]

The World Health Organization had alerted of the disastrous state of the health system in Venezuela (where even hospitals do not have continuous running water and electricity), the poor sanitary and nutrition conditions of 90% of the population, and the reemergence of once eradicated diseases, way before the coronavirus pandemic. Such a scenario of health meltdown looks nowadays not only bleak for the country but also dangerous for the region.[2]

This "Socialist" experience initiated with a hostile attitude towards cities particularly a disdain for large metropolitan areas, in a country with 87% of the urban population, and in a millennium marked by processes of accelerated urbanization and the emergence of mega-cities/mega-regions in many cases driven by informal urbanization. The biases against cities stemmed from the notion that cities embed the evils of class struggles and exploitation, and thus there was an inherent desire to be able to return to a pastoral bon sauvage scenario. An unfortunate outcome was the neglect and deterioration of the urban and territorial arena and therefore for a drastic decrease in living standards for the majority of the population.[3]

The Venezuelan failed experiment illustrates that addressing post-capitalism cannot be reduced to a futile ideological debate between the right and the left, Capitalism or Socialism. It is about finding better ways to foster inclusive societies living predominantly in cities, with environmental responsibility, and making them internally productive with intelligent use of their resources, providing for amicable neighborhoods, infrastructure, services, amenities, and gratifying living conditions. Also, that the territorial and urban, spatial and performative conditions are important variables in this equation. When you have experienced the consequences of ideological dogmatism, deprived of constructive proposals and results, one becomes skeptical of utopian visions, cry for revolutions, and labels. Evolutionary changes build on the preceding conditions, on processes of changes and adaptations. They also respond to global forces but must rely on site-specific responses (see Figures 8.3 and 8.4).

Contemporary urban challenges and informal urbanism

Some of the cities of the Global South are projected to become the largest urban conglomerates in history. At least half of this urbanization process is estimated to occur as self-constructed or informal settlements. Addressing the challenges of the informal city is a task of pivotal importance in light of the demographic tendencies, prolonged life expectancy, the degree of planetary urbanization, food and water scarcity, migratory trends, the rising social tension, and political polarization, and climate change (see Figure 8.5).[4]

Figure 8.3 The two faces of Caracas, the Venezuelan capital, here skyscrapers in the center city.

Photo: Nicola Rocco, Caracas Cenital

Figure 8.4 One-third of the population of Caracas lives in dense informal settlements.
Photo: Nicola Rocco, Caracas Cenital

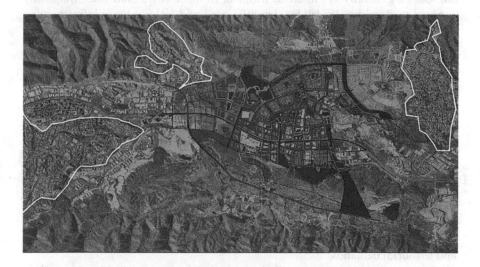

Figure 8.5 Urban design plan for Ciudad Fajardo, Venezuela.
Peripheral informal settlements (in polygons) are excluded from the planned areas, by Luis Sully, Luis Pernía, and Luis Terife, at Universidad Metropolitana, Caracas. Instructor: David Gouverneur
Photo: Luis Sully

Modernist planning paradigms based on zoning and real-estate-driven opera-tions have guided urban growth from the mid-20th century to the present, foster-ing social inequalities, urban dysfunctionalities, and environmental stress. In cities of the Global South the impact has been even more acute. Due to the colonial past of most emerging economies, land tenure remains in very few hands. Thus, landowners have become even richer as rural land is transformed into urban. In turn, the less affluent are pushed – off the map – onto peripheral, under-serviced, inadequate, or high-risk locations, simply because the less-affluent communities cannot access the formal real-estate markets since they lack formal jobs, stable revenues, or savings.[5]

Consequently, a high percentage of the population has no other option than to squat on land and self-construct their dwellings and neighborhoods. Precarious in early stages, they quickly evolve and become part of very large urban areas. Informal urbanism embeds positive aspects such as the velocity in which they emerge and transform providing shelter, the compact use of land, low energy consumption, pedestrian-friendly environments, mixed-uses, a strong sense of place and social ties, all conditions that would be deemed desirable for a sustainable city. In order words, informal urbanism is the only solution for a vast majority, a process that capital-driven urbanism cannot address and certainly not at the same speed. While informality operates avoiding the red tape of formal planning and regulations, both forms or urbanization and economies are intertwined.[6]

But informal urbanism is not exempt from severe problems. The settlements are located generally at the urban fringe or in inner-city's unfit sites, frequently exposed to flooding, landslides, landfills, etc., and lacking basic infrastructure, public spaces, and amenities. In many countries, inaccessibility deprives these neighborhoods from transportation, access to service vehicles (waste retrieval, ambulances, fire-fighters, and police surveillance), making them a haven for drug-trade, gangs, and violence.[7] In some countries, governments evict them after years of communal efforts, piecemeal investments, and social networking.

Although the settlements are of an incremental nature and do not rely upon heavy technologies as large earth moves, they gradually affect environmental systems, degrading areas of biodiversity, polluting and segmenting the hydrologic networks, occupying rich agricultural lands and causing erosion and land instability. Furthermore, they remain in a submissive condition to the formal city,[8] since most jobs and services and amenities remain in the formal city, forcing residents of the informal city to commute. Under these conditions, the entire urban system (the formal and the informal) is strained. It is also important to note that informality works better at a neighborhood scale, but has difficulties responding to urban and territorial demands.

Most communities can gradually improve their homes, starting from precarious shelters they gradually consolidate and expand tailoring them to their needs, and also become income-generating assets by incorporating commercial activities or rental units. Thus, in the informal city, housing is not the main challenge, rather the lack of urban frameworks that can provide for safe and amicable habitats (see Figure 8.6 and 8.7).

Acknowledging this fact, towards the early 1980s, successful programs were envisioned that facilitated self-constructed housing known as "Site and Services." These were first advanced in Academia and later implemented with the support of Governments and Global institutions.[10] The programs provided land in

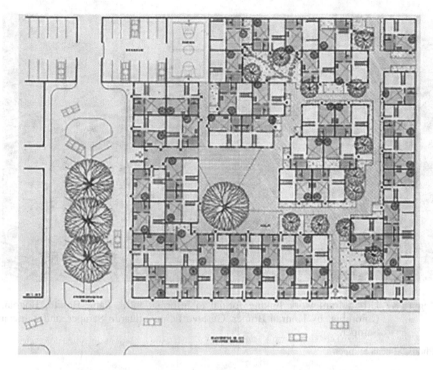

Figure 8.6 Board of a Site and Services project for the PREVI competition, Lima, Peru, by Germán Samper.[9]

Photo: Archives of Germán Samper

appropriate locations, defined the urban grids, the location of future communal services, allocated lots for the construction of homes, and offered families technical assistance. Such programs worked also well at a community scale but did not address the urban nor the territorial scales. Neither did they consider aspects crucial of contemporary urbanism such as climate change, food scarcity, alternative economies, green infrastructure or green energy, migrations, etc. These initiatives fell off the academic and institutional radars by the 1990s. However, there are important lessons to be learned from them, useful for envisioning how to guide larger self-constructed urban territories and meet contemporary challenges.

Informal Armatures: a hybrid approach

We require new and effective urban paradigms for more inclusive and environmentally responsible cities in order to accommodate over one billion settlers that will occupy new self-constructed urban scenarios during the next two decades. How can we counteract the negative effects of conventional city planning and current real-estate forces? How can we boost the potentials of these new settlements

Figure 8.7 Urban framework linking formal to informal infill, Patio Bonito, Bogotá, Colombia, by Konrad Bruner, Gustavo Perry, Eduardo Samper, and Ximena Samper.

Photo: Ximena Samper.

and address their deficiencies? How can we establish synergies between the formal and the informal areas?

I propose that an intermediate or hybrid condition can be achieved by taking advantage of informal dynamics and addressing their weaknesses, combined with creative planning, design, and management. The goal is to homologate living conditions between the formal and the informal city. It can be considered an alternative model of community-based urbanization, offering them the possibility to shape their habitats tailored to their needs, freeing them from the burden of real-estate forces, banking systems, and insurance companies, while attending the aspects that the settlements cannot achieve on their own.[11]

This approach previsualizes informal occupation before it occurs and accompanies the transformation process, supporting the self-constructed and communal drive. The efforts are centered on environmental aspects, on modeling a flexible/transformative public realm, and on the complexities of large urban and territorial systems that surpass the capabilities of the communities. The approach demands holistic visions, technical and managerial skills, the involvement of the community and a proactive public sector. I have called it the Informal Armatures approach, which is described in my publication Planning and Design for Future Informal Settlements: Shaping the Self-Constructed City[12] or its expanded and updated version Spanish: *Diseño de Nuevos Asentamientos Informales*.[13] The following are pivotal aspects of such a hybrid approach.

Anti-bias, political and professional support

For decades, in most countries of the Global South, there was a strong bias against the self-constructed city, demonizing informality as illegal, undesirable, backward, violent/dangerous and consequently targeted for elimination. This bias has deep cultural roots within highly segregated and unequal societies which can be traced to their colonial past. But it also stems from academic and professional misconceptions concerning their role and responsibilities concerning planning, design, and management of cities. Let us remember that modern planning considered the traditional-organic city (as the informal city is) intrinsically sick, and thus required to be eradicated.

Thus, in most developing countries when the first informal settlements of the industrial and modern era began to emerge as a result of rural to urban migration, academics, professionals, politicians, and those living in the formal city believed that informality could be stopped and eradicated. But when the informal city became important, and many times the dominant form of urbanization, and as their residents gained political influence, they started being attended, first in a simple manner as providing them with basic utilities, paving pedestrian paths and stairs, small educational and sports facilities, etc. In time, legal reforms were introduced in many nations to officially recognize informality as a component of city-making and demanding the public sector to act upon them (see Figures 8.8–8.10).

Figure 8.8 Metro-cable station and urban improvements. Barrio Santo Domingo, Medellín.
Photo: Alejandro Echeverri.

Figure 8.9 Pedestrian bridge over a ravine and a recreational pod, Barrio Santo Domingo, Medellín.

Photo: Alejandro Echeverri.

Figure 8.10 Community plaza, Barrio Santo Domingo, Medellín.

Photo: Alejandro Echeverri.

Important initiatives geared to the improvement of existing informal settlements have been carried during the past two decades, particularly in Latin America, demonstrating the value of combining the dynamism of informality with cutting edge planning, design, and management. The Colombian city of Medellín is perhaps the best example of the positive impact of such surgical interventions if they are undertaken as holistic operations with high-quality technical standards and proactive political support. The radical transformations of the once most challenged communities of this city, which was considered during the 1980s the most violent urban center in the world, in terms of livability, community self-esteem, productivity, economic drive, and the nature of the urban, landscape architecture, and architectural interventions have made Medellín an obliged urban and social reference.[14]

The outstanding technical in management and design, as well as of communal engagement, and transparency were also aspects that facilitated this degree of urban transformation. The participation of highly qualified professionals and the steady cash flow from *La Empresa Pública de Medellín* or EPM (the Municipal Holding for all utilities) contributed to the success story.

Such initiatives demanded carefully orchestrated and phased on-site efforts. Partial substitution housing was also required to make room for infrastructure, communal services, recreational and sports facilities, open spaces, or to relocate those settled on high-risk areas. This is a difficult task since the sites usually have no vacant land and nobody wants to move away from their neighborhoods, nor would accept to be relocated unless by doing so they are significantly improving their living conditions, but it is frequently the initial hurdle to overcome to allow for these projects to move ahead. The Medellín case study has not been surpassed in impact or emulated with a similar degree of success in other contexts, probably because of less political commitment and creativity if compared to the efforts of the administration of Mayor Sergio Fajardo – who initiated the transformations – and by those that followed him.

There are still strong biases against informality, and lack of political will to engage such initiatives that require hard work, dedication, that fall under the scrutiny of the communities and thus are less prone to scams and corruption. But, nowadays general academia, professional practice, and government understand the value of addressing the improvement of the informal city. However, introducing urban frameworks into consolidated informal areas is complex, time-consuming, and also costly. There are also limitations of what can be achieved concerning the scale and nature of the projects, without disturbing the settlements, due to the tight urban patterns and many times difficult site conditions.

However, planning for the emergence of new informal areas to foster the growth of the informal city is still considered taboo, illegal, and inconvenient, since it represents a threat to landowners, to the real-estate business, to the bureaucracies and the cultural values of the elites. Conversely, letting the unplanned growth of the informal city to occur at the predicted unprecedented rate, looking the other way and trying to resolve the problem decades from now, will derive in severe social and environmental stress. The key to making a real difference is certainly gaining political support to assist the emergence and performance of the self-constructed city, recognizing the enormous potential embedded in community-driven city-making, accompanied by a proactive and knowledgeable public sector, NGOs, and even the private sector, and yes capital-driven forces, as the settlements evolve. All these arguments favor the Informal Armatures approach, as a preemptive method to assist the emergence and transformation of new settlements.

Land-banking and public-realm armatures

A powerful tool to counteract the exclusionary and environmentally damaging effects of real- estate-driven urbanism is for the public sector to assemble land, enabling the communities to self-construct their dwellings and neighborhoods in safe locations and under suitable conditions. Land-banking should be done preferably before the urban plans are enacted, in anticipation of the increased price of the land, and capturing created real-estate value for the public good. The underlying premise is that land to dwell, to derive food from, and to provide ecological services should be a fundamental human right. Land assemblage can be considered an upfront subsidy that will avoid future social tension, loss of lives and goods. Assembling public land – before informal occupation occurs – also allows for a more equitable spatial distribution of the different income groups.

The approach emulates and surpasses the benefits of the "surgery projects" in informal settlements, acting preemptively and deploying at different scales the environmental and urban support systems. The political, design, financial, and managerial efforts of such armatures should be geared to shaping the public realm and adapting it to the changing needs of the communities as they become part of large urban and territorial systems. It is important to mention that, under non-fostered conditions, informal urbanization leaves no room for public spaces or communal services, neither does it protect the environmental systems.

Thus, identifying the areas to be protected, as well as defining the systems of open spaces (enabling the flow of people, goods, water, economies, etc.) by modeling these territorial armatures and monitoring their transformations is an important task. Equally relevant is keeping these armatures free from unwanted occupation. In the approach, the construction of mixed-use and expandable dwellings continues to be the responsibility of the settlers, while they can benefit from receiving technical and financial support (see Figures 8.11 and 8.12).[15]

Safeguarding natural and cultural landscapes

The aforementioned armatures can be deployed on to be urbanized territories, protecting environmentally sensitive areas, such as watersheds and streams, flood-prone areas, depositories of bio-diversity, good quality soils/agricultural landscapes, archeological sites, etc., all of which tend to be eroded in contemporary driven practices, particularly in developing countries.

In the Informal Armatures approach these environmentally sensitive areas have been designated as Protector Corridors (see Figure 8.13). Safeguarding these

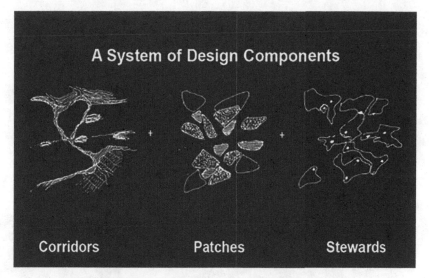

Figure 8.11 A system of design components of the Informal Armatures approach.
Diagram: David Gouverneur

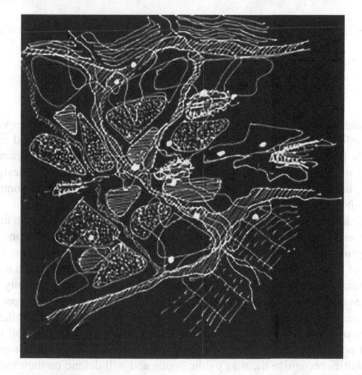

Figure 8.12 Composite of components.
Diagram: David Gouverneur

Figure 8.13a Protector corridors, safeguarding environmental assets with uses relevant to the communities.

Diagram: David Gouverneur

assets can be successfully achieved in this context if such moves are envisioned before informal urban occupation takes place. Green infrastructure and environmentally friendly practices can be best embedded in the systems working with the communities from the early phases of occupation. The planning, design, and managerial efforts should be geared to ease development pressures onto these "green sites."[16]

A viable way to protect these sensitive sites is to associate them with uses that will be meaningful to the community, e.g., food gardens or recreational areas, accompanied by respected institutions that can serve as stewards, thus keeping them free from informal occupation. The Informal Armatures approach also suggests associating these environmental assets with respected community "stewards." Such stewards can act as "garden keepers" of these open spaces and communal services, carrying out projects and activities with the dwellers. The territory under the custody of the stewards may require to be fenced off or have limited access until the activities and projects are carried out and or until the communities recognize them as public goods and will defend on their own from unwanted forms of informal or formal occupation. These activities may also change during the different phases of evolution of the settlements, responding to the new community and urban demands.

Figure 8.13b Recreational area adjacent to the Catuche Ravine, Caracas.

Photo: David Gouverneur

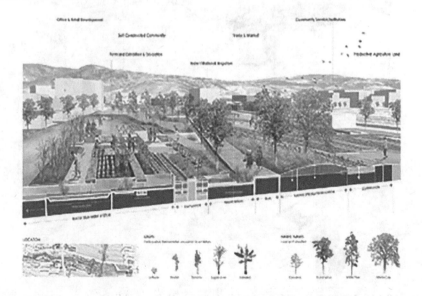

Figure 8.13 c Community vegetable garden for Villanueva, Ciudad de Guatemala.

Project and Photo by Zihan Zhu and Xu Han, Instructor: David Gouverneur, Guatemala Studio, Weitzman School of Design, University of Pennsylvania

Configuring attractor, connectivity, and service corridors

The predominantly self-constructed cities are by nature compact, low carbon impact, and pedestrian-friendly (the opposite of the "other" city). They also tend to be dense, lively, of a mixed-use nature, and pedestrian-friendly, comprising a continuous and homogeneous quilt of neighborhoods. What is lacking is better connectivity and mobility systems for pedestrians, bikes and public transportation, public spaces, stronger and diverse economic drivers and productivity, basic communal services as markets, hospitals, libraries, techno-parks, universities, art centers, sites of larger civic encounter for festivals, rituals, sports, communal facilities for stronger political interplay and self-governance, etc., which tend to be located only in the formal city.

As in the case of the Protector Corridors, the Informal Armatures approach envisions pre-defining the location and the basic morphology of the Activity Corridors (see Figure 8.14). Concentrating design and managerial efforts in these

Figure 8.14a Attractor corridor providing connectivity, public spaces, services, and activities.

Diagram: David Gouverneur

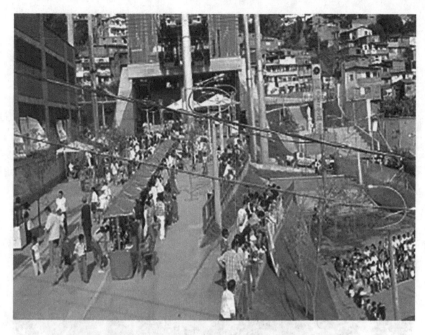

Figure 8.14b Informal market adjacent to a Metrocable station, Barrio Santo Domingo, Medellín.

Photo: Alejandro Echeverri

areas allow balancing the provision of such services and amenities and infra-structure throughout the entire urban system, reducing inequalities. It also allows diminishing pressures onto the green corridors. These higher-energy Attractor Corridors are expected to also transform over time, as the districts grow in popu-lation, become more diverse, complex, as the areas increase in population, expand and also as communal expectations rise.

The section of the Corridor should be designed to be able to accommodate changes and including Stewards may facilitate these changes. Let us imagine that a particular Attractor Corridor is the desired location to introduce a public transit system in the future, serving the adjacent areas and civic amenities; but this trans-portation system cannot be introduced until there is critical population mass. The Attractor Corridor could begin by incorporating bike and pedestrian lanes, while part of it could be designated for fenced community gardens under the stewardship of a respected institution or communal organization. This organization can pro-vide food, potable water, or planting of wood trees for the construction of homes in the initial phases of occupation. Over time, these activities can be transferred to new sites assisting new emergent occupations, and the Corridor can be retro-fitted to accommodate the public mobility systems, civic spaces, and associated

Figure 8.14c Library and cultural activities in an informal settlement of Medellín.

Photo: Alejandro Echeverri

communal services. In both cases, the Protector and Attractor Corridors combine territorial-physical and performative conditions, acting as a framework supporting the areas designated for the self-construction of dwellings/neighborhoods, which have been designated as Receptor Patches, as well as for other supplementary uses at a local, urban, or territorial scales, identified as the Productive Patches.

Receptor patches and productive/income-generating patches

Informal occupation always departs from occupying a plot of land. This provides the residents with a sense of ownership, even in unplanned settlements where the possession of land is initially considered illegal, or when it may take years before land-titles are granted. Thus, in most countries of the Global South, informal occupation is there to stay, unless they have been located in high-risk terrain, or when relocation is required to introduce improvements that will contribute to a greater common good, including benefits to those displaced.

The Informal Armatures approach offers the dwellers risk-free land, suitably located, supported by the aforementioned Corridors, to be developed in the Receptor Patches. If such availability of land is not provided, informality will continue to emerge in inappropriate and unserved sites, continuing the cycle of exclusion, inequity and making the cities even more dysfunctional, particularly in very large informally urbanized territories. According to the technical and managerial resources of each city/district, the Receptor Patches may evolve as they have spontaneously done by individual or communal efforts or can be subject to Site and Services plans, predefining the urban grids, public spaces, lots/basic shelter and infrastructure, communal gardens, etc. In both cases, the public or communal facilitators of the operations, working with the community leaders, can establish mechanisms to ensure a fair allocation of the lots to avoid speculation, assign land titles, safeguard the land set aside for communal uses, provide technical assistance and maintenance, etc. (see Figures 8.15–8.17).

Residents can gradually enlarge their dwellings, accommodating for demands of the growing or expanded families, allowing them to rent or sell sections of the enlarged homes. Furthermore, it is possible to introduce in the dwellings productive/income-generating uses, as small shops, restaurants, bodegas, manufacturing,

Figure 8.15 Receptor Patches (in black dots) to facilitate self-constructed neighborhoods.
Diagram: David Gouverneur

Figure 8.16 Consolidated barrio in Chacao, Caracas.

Photo: David Gouverneur

Figure 8.17 Site and services drill at Upenn by students of the Weiztman School of Design,
         University of Pennsylvania campus.

Instructors: David Gouverneur and David Maestres

Photos: David Gouverneur

repair services, etc. These uses foster entrepreneurship and the maintenance and constant improvements of the building stock. Rarely these conditions are present in formal urbanism/real-estate-driven operations, particularly in social housing programs.

However, the better-paid jobs and urban services are located in the formal city, forcing the settlers to commute long distances. The flexible micro-economies are important to the neighborhoods but have a lesser impact and aggregated value than the economy of the formal city. The Informal Armatures approach seeks to boost the economies within the informal city, establishing synergies with the formal modes of production.

A combination of territorial/spatial and performative/managerial moves can determine the location and dimensions of Productive Patches, allowing to gradually incorporate larger-scale manufacturing, artisanal, commercial or manufacturing uses enhancing the capabilities of the micro-economic activities carried out in the dwellings (see Figs. 8.18–8.20). For instance, in south Lima, Peru, a two-hectare facility was developed with the support of an NGO. It serves as a window-front and marketing center for hundreds of small businesses with products that are manufactured in nearby informal communities. Here, clients from all over the

Figure 8.18 Productive Patches (in red dots) to ensure supplementary urban uses.

Diagram: David Gouverneur

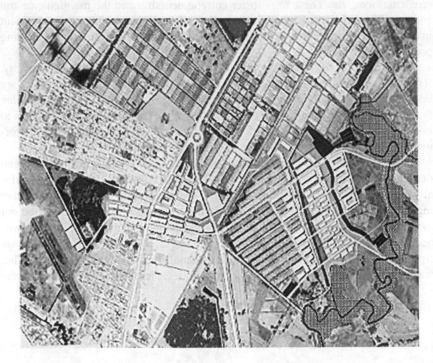

Figure 8.19 Green houses for flower production, Sabana de Bogotá Studio, School of
Design, University of Pennsylvania.

Project: Tamara Henry, Instructor: David Gouverneur

country and from abroad visit this center or contact it to order larger quantities
of high-quality and competitively priced and locally made items (furniture and
fixtures, mechanical appliances, textiles, lamps, commercial refrigerators, etc.).
This type of cooperative organization and manufacturing incubators can ensure
client accessibility, visibility/marketing, the rate of supply, reliability, commercial
patents, and a healthy correlation of labor–income, all conditions that could not be
achieved by the small entrepreneurs working on their own.

Over time, some of the Productive Patches can be used to attract investors from
the formal sector to build and/or to operate projects, allowing for diversification of
uses and the mix of income groups. The private sector is usually willing to invest
in these districts when the population and their income and aspirations augment,
for instance in cinema, commercial, or recreational complexes. But land to do this
is usually not available within the dense non-fostered informal settlements. The
benefit of this modality is that the plus-value of the urban land can be captured
by the public sector that envisioned the Informal Armatures district to start with,
and not by private landowners as usually would occur under the conventional

Figure 8.20 A community center and park over a defunct quarry, Sabana de Bogotá Studio, School of Design, University of Pennsylvania.

Project: Rachel Ahern, Instructor: David Gouverneur

zoning-based approach. The capital gain of these operations can be used for new investments in the district or to advance similar programs in new territories.

Another modality advantage of "Productive Patches" is that they can be used for larger-scale community-based services like parks, schools, technical schools, universities, hospitals, etc. when the demographics demand them. The funding to develop these public uses can be embedded in the zoning codes of the formal city, where the right to urbanize private land requires the developers to include open spaces and communal services on-site, and/or to pay taxes/urban development fees. This option

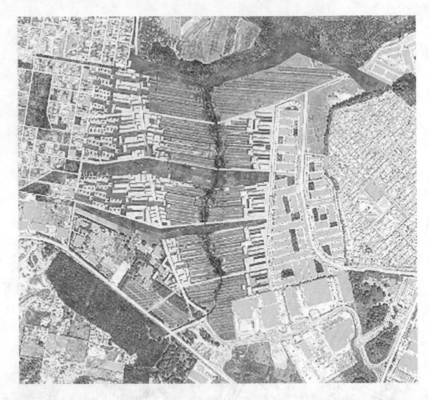

Figure 8.21 Agricultural district for the peripheral municipality of Villanueva, Ciudad de Guatemala, Guatemala Studio, Weitzman School of Design, University of Pennsylvania.

Project: Zihan Zhu and Xu Han

Instructor: David Gouverneur

works as a cross-subsidy from real-estate-driven areas to the predominantly self-constructed areas. The public sector should be vigilant to ensure that the quality of these community interventions serving the self-constructed neighborhoods is equal to that normally achieved within the real-estate-driven operations. There could also be innovative modes of operating these facilities as public–private ventures, as has occurred with the new educational facilities constructed in challenged informal neighborhoods during the first mandate of Mayor Enrique Peñalosa in Bogotá.

Fostering community-based agriculture

Food sufficiency is a major concern in the immediate future until radically new forms of production are achieved. Agriculture can be produced within the Informal Armatures districts at different scales and moments, whether in the early stages of the Corridors or Productive Patches, or within the Receptor Patches (at a family or communal scale). This could be geared not only to offer good quality products at low prices for the residents of these emergent districts (avoiding transportation costs) but also to provide a surplus to sell the agricultural products and derivates to the formal city. Community-based agriculture has a better chance to prosper in these emergent areas since the Informal Armatures approach is usually deployed in peri-urban sites, safeguarding valuable agricultural land. Additionally, a high percentage of new settlers usually come from rural areas and have agricultural and grassing skills. These abilities are quickly lost in the non-fostered informal settlements.

The Informal Armatures approach may also be useful in establishing spatial and productive links between rural communities adjacent to the urban and metropolitan areas. This aspect has not been addressed by conventional planning. A goal here is to protect and boost rural areas located from being absorbed by urban growth, enhancing agricultural productivity, increasing revenues for these communities, and providing them with services that are usually only available in the urban context. Additional benefits are the protection of watersheds and flood-prone areas, as well as safeguarding cultural practices, culinary, festivals, crafts, and forms of agrotourism in the adjacencies of highly urbanized areas. In this hybrid vision, both rural and urban would benefit (Figures 8.21-8.22).

Co-participatory planning, design, execution, management, and governance

The aforementioned strategies can be best achieved by establishing synergy among the communities, the public sector facilitators, NGOs, institutions, etc. The teams of facilitators of the approach require an ample array of cross-disciplinary skills, coupled with the ability to engage the communities, and establish trust, leading to collaborative work, especially until the communities can take over. This requires proactive actions on the site and not remote office-work delivering detached and top-down decisions, which has been common in most cases in conventional planning and management.

A difficult role of such facilitators is to respond to the immediate demands of the settlers in the early phases of occupation, while not losing sight of future urban and territorial demands, as the settlements become part of much larger and complex territories. For instance, early settlers are mostly concerned with securing a lot and gathering material to initiate the construction of their shelters/seeds of their future homes, gaining access to food and potable water, avoiding risks and violence, and begin navigating urbanity. At this early stage, it is difficult for them to envision that in a few years or decades they will be part of much larger conglomerates that require more sophisticated support systems, e.g., services, amenities, public transportation, parks, as well as large urban and metropolitan facilities. It is also difficult for the community to envision that their standards of living will be greatly conditioned by safeguarding the environmental assets of the territories they are occupying.

Since each context presents different challenges: lack of potable water, lands instability, high levels of violence, religious or ethnic conflict, etc. Also, each territory has different potentials: fertile soils, the possibility of tapping solar energy, communal organization, craftsmanship, economic stability, etc. Therefore, counting with expertise in these particular fields is most relevant. Cross-disciplinary work is crucial to maximize the positive impact of the initial occupation and sustain the transformation of the settlements and how they perform as part of large systems. The ultimate goal is for the communities to quickly take over the management/governance of their neighborhoods and districts, allowing the public sector to gear similar efforts in the initial phases of occupation of emergent settlements, and focus on larger-scale systemic aspects.

The facilitators of the Informal Armatures approach will have to deal with politicians, institutions, and above all, with the communities, thus political abilities, social and communication skills are equally important. However, lower-income communities of the Global South are skeptical of the managerial skills and the honesty of the politicians/public sector. Corruption is always stated as the forerunner of socio-economic inequality. Training these versatile and honest facilitators is a critical factor to make the Informal Armatures approach viable. Such an agenda requires new technical, political, and managerial skills. We are talking of a decentralized, more democratic, transparent, and participatory approach of city-making.

We are in urgent need of alternative planning, design, and managerial paradigms that are centered on the welfare of the majority of the population with a respectful attitude towards the environment and the natural and cultural resources. This alternative planning, design, and managerial approach should be far from the nature and scale of formal/speculative urban operations that have been favored throughout the developing world. Their outcome has been negative environmental,

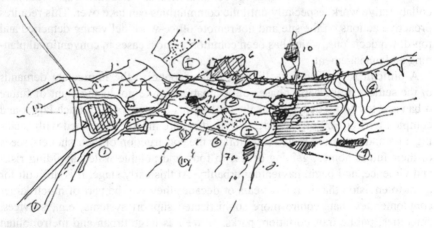

Figure 8.22 Informal Armatures.

Sketch: David Gouverneur

social and economic effects, many of which have been carried out by international corporations/capital in agreement with the political, financial, and technical elites in these countries.

Taking into consideration the sheer demographic projections, and the global impact of the self-constructed city during the following decades, the Informal Armatures approach seems to be a viable option, aimed at tapping the creativity, energy, and self-determination of the communities. It seeks to establish synergies among complex environmental, urban, and socio-economic and cultural aspects, providing support at multiple scales, and adapting to changing conditions. This approach can also help to improve the performance of the formal city when conceived as an integrated and hybrid urban system. There is not such a thing as a post-capitalist city or a socialist city. The city of the future will not result from a monolithic utopia; rather from a more diverse, flexible, inclusive, economically fluid, greener, energy-efficient, and community-governed city. Addressing the demands of the predominantly self-constructed city –where one-third of the planet's population will reside – may help us move in this direction.

Notes

1 See BBC news (Jan, 13, 2020), Venezuela crisis: How the political situation escalated-www.bbc.com/news/world-latin-america-36319877
2 As of early June 2020, only 18 deaths had been reported; some argue thanks to the isolation of the country with virtually no international flights before the pandemic and no tourism, limited mobility within regions due to the lack of gasoline, repressive measure to enforce the lockup, and governmental control of the information.
3 See article Misery stalks Venezuela. Report from the Miami Herald. March 27, 2018www.miamiherald.com/news/nation-world/world/americas/venezuela/article206950449.html.
4 See United Nations, DESA, Population Division (2005). World Urbanization ProspectsThe 2005 Revision. UN Report, United Nations Publications, New York, USA.
5 D. Gouverneur (2019) A Mosaic of Enhanced influences, Landscape Architecture Journal (with Chinese translation by Xian Luanli, Denk Ke), The Latin American Landscape Journal, Beijing, China.
6 De Soto, Hernando (1989), The Other Path, J: Harper & Row Publishers, Inc. New York, USA.
7 For detailed statistic on sources and type of violence in the slums see United Nations Habitat. The challenge of slums. Global report on Human Settlements. Official, London: Earthscan Publications Ltd, 2003
8 For issues derived from urban inequalities revise Grauer, Oscar. Democracy and the City, in Democracy in Latin America, ReVista. David Rockefeller Center for Latin American Studies, Harvard University, Fall 200
9 The housing stock may be at risk of flooding, hurricanes, landslides, fires, vulcanism, and quakes or earthquakes. These aspects may derive from inadequate site-conditions and/or technical limitations or lack of proper technical assistance.
10 See John F. C. Turner y Robert Fichter (1972), Freedom to Build: Dweller Control of the Housing Process, Macmillan, New York, USA.
11 See Keri E. Iyall Smith and Patricia Leavy (2008), Hybrid Identities, Brill, USA.
12 Gouverneur, D (2014b). Planning and Design for Future Informal Settlements: Shaping the Self-Constructed City, Routledge, Oxford, UK.

13 Gouverneur, D. (2016). Diseño de Nuevos Asentamientos Informales, Fondo Editorial de la Universidad de La Salle, Bogotá y Fondo Editorial de EAFIT, Medellín, Colombia, 2016

14 Medellín received the "Innovative City of the Year" award, in 2013 granted by *The Wall Street Journal* from a shortlist that included New York City and Tel-Aviv. This was one of many prizes and distinctions that Medellín has received, a city which was considered for decades the most violent urban center in the world. Medellín exemplifies that even under the most adverse conditions urban improvement is possible, and in a short time, it became a reference for many developing nations and particularly in relation to the improvement of urban conditions.

15 See Chapter: A System of Components in Gouverneur, D. (2014a), Planning and Design for Future Informal Settlements: Shaping the Self-Constructed City, Routledge, Oxford, UK.

16 M. Rios, (2015) "Marginality and the prospect for urbanism in the post-ecological city" Landscape Urbanism and its Discontents. Dissimulating the Sustainable City. eds. Andrés Duany and Emily Talen, New Society Publishers, BC, Canada.

References

BBC News, 2020, 'Venezuela crisis: How the political situation escalated', 13 January, viewed 4 October 2020, fromwww.bbc.com/news/world-latin-america-36319877.

Freisler, E., 2018, 'In Venezuela, hungry child gangs use machetes to fight for 'quality' garbage', *Miami Herald*, 27 March, viewed 8 October 2020, fromwww.miamiherald.com/news/nation-world/world/americas/venezuela/article206950449.html.

Gouverneur, D.C., 2014a, 'The IA as a system of components guided by principles of implementation', in D. Gouverneur (ed.), *Planning and Design for Future Informal Settlements: Shaping the Self-Constructed City*, pp. 163–202, Taylor and Francis.

Gouverneur, D.C., 2014b, *Planning and design for future informal settlements: Shaping the self-constructed city*, Taylor and Francis, https://doi.org/10.4324/9781315765938.

Gouverneur, D.C., 2016, *Diseño de nuevos asentamientos informales*, Universidad La Salle, Fondo Editorial Universidad EAFIT, Medellín.

Gouverneur, D.C., 2018, 'The Latin American Urban Landscape: A Mosaic of Enhanced Influences', transl. X. Luanli & D. Ke, *Landscape Architecture Frontiers*, Vol. 7, pp. 1–24.

Grauer, O., 2002, 'Democracy and the city: My city, your city . . . one city', *ReVista : Harvard Review of Latin America* 2(1), 16–20.

Rios, M., 2013, " 'Marginality and the Prospect for Urbanism in the Post-Ecological City', in A. Duany & E. Talen (eds.), *Landscape urbanism and its discontents: Dissimulating the sustainable city*, pp. 199–214, New Society Publishers, Gabriola Island, BC.

Smith, I.K.E. & Leavy, P., 2008, *Hybrid Identities: Theoretical and Empirical Examinations*, Vol. 12, Brill, Boston, MA.

Soto, H.de., 2002, *The Other Path: The Economic Answer to Terrorism*, Basic Books, New York, NY.

Turner, J.F.C., 1972, *Freedom to Build; Dweller Control of the Housing Process*, Macmillan, New York, NY.

UN-Habitat, 2003, *The Challenge of Slums*, Earthscan Publications, London ; Sterling, VA.

United Nations and Department of Economic and Social Affairs, Population Division, 2015, *World Urbanization Prospects: The 2015 Revision*, UN Report, United Nations Publications, New York, NY, https://population.un.org/wup/Download/

9   What will a non-capitalist city look like?

Tom Angotti

Introduction

In answering the question of what a non-capitalist city will look we could fall into the trap of physical determinism, the occupational hazard of urban planners, architects and designers. Our challenge is not to paint pretty pictures of the future city without capitalism but to unfold its essential elements, which are economic, social and political – elements that would be reflected in the physical city but would not individually or neatly correspond in any direct or parallel way with them.

The capitalist city has reached the end of its useable lifetime as catastrophic global climate change approaches and we urgently need alternatives. Even if the capitalist city by itself is not the cause of climate change it is a major factor in the global climate crisis. The city today is the site for global capitalism and its unsustainable production of greenhouse gases. Neither the capitalist city nor its host can survive unless there are major changes in the way humans live with the earth and relate to land. The real question is what, if anything, will replace the capitalist city as we know it?

Contrary to the neoliberal meme "There Is No Alternative" that was popularized by British Prime Minister Margaret Thatcher, there *must* be alternatives or we face massive planetary destruction, species loss and the end of the Anthropocene. In the struggles to halt global warming it is urgent to develop real alternatives, not abstract utopias leaping from the imaginations of a few individuals but rather an array of imaginative alternatives that spring from the long traditions of struggle by those who have been marginalized, victimized and displaced during the urbanization of the planet over the last two centuries, those who are most vulnerable to climate change.

Before attempting to define what an alternative, non-capitalist city might look like we should clearly understand the main problems with the capitalist city and why they are unlikely to be resolved in a global economy driven by capitalism. My premise is that the problems of the modern metropolis are intimately linked to its emergence under global capitalism and that its deep structural problems, especially the commodification of land, cannot be resolved under capitalism (Angotti 2012). Any transition towards the non-capitalist city will confront the legacy of centuries of capitalist development.

The cities of global capitalism are at the heart of a world in which the reckless production and consumption of commodities is driving catastrophic climate change. Without visions of a non-capitalist world to drive protest, resistance and deep structural changes, humans will likely hum along as if nothing were amiss, careen over an historic cliff and have no control over where they may land. Visions of a global future with new and different cities can help prevent climate disasters by moving humanity towards a different and better world. In the United States, they might inform the "Green New Deal" and other efforts to drastically reduce the carbon footprint of our urbanized world.

To imagine "what the non-capitalist city will look like" is to construct a broad vision encompassing the physical, social and economic city. A strict physical determinism would limit the discussion to only the visible, physical city and evade discussion of the economic and social relations underlying them. It could also fall into an anthropocentrism that fails to take into account the complex relationships between sentient humans and the full range of other living species on the earth, most of whom are in non-urban habitats. Human activity on the earth is resulting in a drastic and accelerating rate of species loss, and the urban habitat created by humans is anthropocentric to the point of contaminating or eliminating the habitats to which other species have adapted and in which they thrived. This includes the degradation of marine habitats, ocean acidification and the loss of marine habitats. This too is accelerating global climate change which, in turn, threatens the urbanized world.

Capitalism is the problem, not the solution

The leading "solutions" to the climate crisis emanating from the centers of global capital add up to "more capitalism." They are presumed to be more practical and "doable" because they do not challenge the fundamental roots of economic and political power underlying the crisis. Carbon trading is the favorite market solution but so far it has failed to bring about meaningful change. Then there are the technological fixes that promote new and lucrative industries like electric cars, solar panels and wind turbines, all of them subject to competition for capital and customers, and favored by government policies protecting and incentivizing the choices of investors. While conversion to solar and wind are among the most likely strategies to head off the climate crisis, without structural changes in the relations of production, reproduction and consumption they are unlikely to be effective in the long term.

Technological fixes include a full array of palliatives, half-measures and boutique products that leave the basic relations of production intact. Most importantly, market-driven solutions are bound to reinforce the gaping economic and social inequalities at the root of climate injustices that expose those not favored with masses of capital to the greatest vulnerabilities to climate change. The neighborhoods and nations in the global South that are now most vulnerable to extreme weather events, food insecurity and the economic and environmental damage caused by sea-level rise will not get the subsidies, tax breaks, or assistance they need. Instead, well-intentioned reformers will advocate for technological and design solutions that presumably would "raise all boats" without regard to the huge imbalance between the billions of tiny boats that get lost at sea and the gilded Noah's Arks built so the rich can survive the most extreme storms.

As Naomi Klein stated so eloquently, climate change "changes everything" (Klein 2014). Without confronting capitalism's insatiable drive for "growth," its need to generate and sell more commodities, cars and houses; its need to extract more mineral wealth from the earth to feed its engines of production; its over-production, underconsumption and ability to "externalize" the costs of waste and pollution, it cannot sustain the globalized system that runs the world's economy (Baran & Sweezy 1966; Foster 2009; Foster, Clark & York 2010).

Since the beginning of the 21st century, most people in the world have lived in large cities and metropolitan regions. These human settlements produce and consume almost all of the greenhouse gases in the world. More than ever before, rural areas provide food and raw materials for the metropolis while other than human species are corralled and cultivated to serve capitalist enterprise or, worse, face extinction (Angus 2016; Bonneuil & Fressoz 2017; Foster 2009; Foster *et al*. 2010).

Marx and Engels recognized that the separation of city and countryside during the industrial revolution was one of the main negative consequences of the rise of capitalism (Marx & Engels 1947; Lefebvre 2016). Following a century and a half of extractive capitalism's destructive drives for global expansion, cities and metropolitan regions are now the uncontested centers of economic and political power and unbridled consumption, while what is left of the rural world continues to be drained of resources or shorn of its natural ability to absorb excess carbon emissions (Angotti 2012).

If global capitalism built the metropolis by plundering rural areas for labor and commodities then the non-capitalist city of the future must drastically reduce the urban–rural divide by helping to repair the ecological rift that was opened up by a century and a half of unremitting capitalist development (Foster *et al*. 2010). Since global capitalism thrives on the endless expansion of consumption to sustain profits, the infrastructure of the modern metropolis is built in order to maximize the exploitation of labor, the unequal reproduction of the labor force, and inefficient individualized consumption. This means, concretely:

- private cars instead of mass transit
- private homes instead of social housing
- conspicuous consumption of mass-produced commodities by individuals, including personal computers, phones and entertainment devices
- the exploitation of women, immigrants and people of color to enable the low-cost reproduction of the labor force
- planned obsolescence, endless cycles of consumption and profits.

The modern metropolis requires vast energy use and accumulates mountains of waste that contaminate air, water and soil, destroy habitat and species and expand inequalities as the economically disadvantaged find themselves living in the most precarious and contaminated environments (Strasser 1999).

Colonialism and slavery are imbedded in the capitalist city

To both imagine and struggle for a non-capitalist city, it is not enough to simply understand and erase the wasteful and destructive characteristics of the modern metropolis. Capitalism's global hegemony would not have been possible without the establishment of colonies and their eventual transformation into dependent capitalist nations. Colonialism gave birth to and fed upon slavery. In today's globalized capitalist world, colonialism and slavery are portrayed as relics of the past but they are very much present behind the sleek veneer of the modern metropolis. They are visible in segregated districts and immigrant enclaves and they thrive with the maintenance and growth of racist and anti-immigrant violence.

Elites and the tourist industry in New York City, for example, project an image of the city as a model of multi-racial and inter-ethnic tolerance and harmony, a glorious "melting pot." The city has immigrant communities from every corner of the world and they live in close proximity to one another. However, beneath the surface of apparent calm lies a different reality. This is one of the most highly segregated cities in the world – by race and ethnicity (Angotti 2017). People of color are more likely to end up in prison because of racist policing and the inability to post cash bail. Their children are more likely to attend overcrowded and under-performing schools. They live in neighborhoods with disproportionately high exposures to toxic wastes (Sze 2007). They do not have access to affordable healthy food and therefore end up sicker. Contrary to the official myths, poor immigrants from poor countries live in poor neighborhoods with people from the same poor countries and regions of the world. The rich and middle classes, on the other hand, live in their upscale bubbles, in both vertical and horizontal enclaves where land values, rents and public services are as elevated as the buildings.

The most powerful capitalist producers in the most powerful nations maintain hegemony over the "post-colonial" world of dependent nations, whose land, producers and workers are unable to break away from a system that produces environmental disasters and destroys livelihoods. They are joining the global migrant streams filling the cities of the global North with refugees who have limited economic opportunities and rights. They are the climate refugees and victims of capitalism's environmental injustices no less than the farmers in 19th-century Europe who fled the enclosures to find work in the early cities of industrial capitalism.

Environmental justice and green gentrification

The denizens of the modern metropolis include legions of dispossessed indigenous people, rural and small-town residents, and climate refugees fleeing from land grabbers, mining monopolies and purveyors of industrial agriculture. In the metropolis they face new and different threats of dispossession. They live at the margins, with low-paid and precarious jobs, in marginal housing and neighborhoods, with environmental and public health hazards, but after struggling to sustain and improve their communities they face rising land and housing values they cannot afford and that push them out. Civic improvements they may have fought hard for may trigger gentrification and eventual displacement to yet another peripheral location. "Green gentrification" can therefore reproduce the global environmental injustices inherent in the capitalist city (Gould & Lewis 2017).

"Green cities," "smart cities," and any number of other trendy notions often put a luster on the expanding inequalities in the world's cities and help to sell real estate at enormously inflated prices. LEED certified "green" buildings add value to luxury buildings whose values are already overinflated by speculative land prices and that are grossly inefficient when considering the amount of space and energy they require per capita. Green roofs grow expensive culinary herbs for gourmet kitchens. Public open space is added as an amenity to boost nearby housing prices, but large open spaces face opposition when they deprive speculators of golden opportunities for investment. Of course, PPPs (Public Private Partnerships) in parks, also known as conservancies, can insure that no green is allowed to grow without somebody making a profit, thereby converting "green" from a use-value to an exchange-value and reproducing capitalism's fundamental problem with nature.

These criticisms are not simply ideological disagreements over the merits of capitalism or neoliberalism but exposing them is part of the larger political struggle over the future of humans on the earth. We are struggling for the survival of the Anthropocene and asking whether it is possible and under what conditions. What should human habitats look like in the new world? Will they be smaller cities and urbanized regions or larger ones? How will humans treat urban and rural land? Will land continue to be a symbol of colonial conquest? Will it be partitioned to preserve racial divides? Is it a commodity to be bought and sold? The same questions arise when dealing with other basic elements like water, air and fire. Most importantly, however, how will humans in our post-capitalist cities deal with each other across class, racial, ethnic, gender and many other divides that characterize the globalized capitalist world and its metropolises?

As part of the political struggle for alternatives, we should outline alternatives both to develop concrete political strategies and to help get them implemented. As we struggle for more sustainable and just ways for humans to live on the planet, we can explore urban and rural utopias that inspire us to act boldly and wisely.

Why we need many utopias

In a speech to the People's University, Arundhati Roy (2014:93) beautifully described the connection between protest, organizing and the pursuit of social justice:

Yesterday morning the police cleared Zuccotti Park, but today the people are back. The police should know that this protest is not a battle for territory.

We're not fighting for the right to occupy a park here or there. We are fighting for Justice. Justice, not just for the people of the United States, but for everybody. What you have achieved since September 17, when the Occupy Movement began in the United States, is to introduce a new imagination, a new political language, into the heart of Empire. You have reintroduced the right to dream into a system that tried to turn everybody into zombies mesmerized into equating mindless consumerism with happiness and fulfillment.

Returning to the question of what a non-capitalist city would look like, we confront the tradition of utopian thinking and its role in the history of urban planning. In response to the failings of capitalism's earliest cities in Europe and the Americas, two grand city planning ideas occupied center stage: the City Beautiful Movement, which emphasized building modern city centers as centers for civic activities and symbols of civic pride; and the Garden City Movement, which sought to create more human-scale new communities providing ample economic and recreational opportunities for all. Both of these led to partial changes in both the physical city and its social and economic base. Both reproduced a dominant male, white proclivity for spatial order and left aside bigger questions of racial, ethnic and gender diversity and economic and social justice.

Within the United States, two major new ideas emerged in the 20th century. The first, an updated version of City Beautiful, charted the reconstruction of dense urban centers as nuclei for expanding metropolitan areas. One of the foremost practitioners was Robert Moses who rebuilt New York City as the triumphant center of a vast suburbanized metropolitan region (Caro 1975). In response to Moses, Jane Jacobs (1961) joined community resistance to displacement and advocated for a more organic, gradual process of urban transformation. In the metropolis of the 21st century both of these tendencies fall short of the challenge posed by climate change. Neither addresses in any comprehensive historical or socio-economic context the relationship between people and land. Caro left us with a city of enclaves while Jacobs, in one of her later books, doubled down on a cultural argument, forecasting "Dark Ages Ahead" due to deep-rooted anti-urbanism (Jacobs 2004).

However, in recent decades a virtually unbounded global real-estate market driven by enormous speculative profits has flourished under neoliberal principles and altered the face of most metropolitan regions around the world. This has dwarfed the relevance of any strong urban planning, planned alternatives and utopias.

I cannot outline in detail what the non-capitalist city will *look* like because our alternative must be to change both the physical, visible city *and* change the means through which the city changes and develops. Change is a dynamic process that will be filled with contradictions (as per dialectics), grounded in real history (as per historical materialism) and class struggle, which will not evaporate in a post-capitalist society but should be transformed into peaceful processes. It is not simply a question of physical planning, design, or cultural tradition – though all of these are important factors. If our post-capitalist world is to be more just and sustainable there cannot be a single vision of the future city such as those that governed colonial and imperial urbanism from ancient Greece to modern South Africa. This is fundamentally a question of political economy not urban design.

Our utopias are not dreams but *political strategies* that can energize the global movements for economic, urban, rural and climate justice. Our visions of the just, non-capitalist city should be many since that city will be made up of diverse people who will have different experiences and views of what a post-capitalist city should look like (see Angotti 2008). Classical utopias by Sir Thomas More and Edward Bellamy, for example, reflected the world views and politics of privileged white men. More had no problem incorporating slaves in his puritanical utopia and leaving the cooking to the women (More 2011; Bellamy 1960). The 19th-century utopian socialists reflected the realities and ideologies emerging from worker dissatisfaction with Europe's industrial capitalism, but they also appealed to isolated strata of the working class. They did not survive very long in a global environment in which capitalist industry overwhelmingly shaped cities and the ways that workers lived.

For our current utopias we might learn much from the inclusionary reflections of science fiction writers like Ursula LeGuin (1969) and story tellers like Toni Morrison (1987) whose imaginations helped transport our minds out of the oppressive conditions that have imprisoned our bodies and minds. These revolutionary thinkers can help us dream and leap into better futures, building bridges over the oppressions of the past and present. These are not simply cultural arguments for new urban futures; they are raw material for liberating ourselves from the prisons of capital's ideological hegemony.

Reinventing relations between time and space

In planning the post-capitalist city, we have much to learn from the many societies in the world in which the boundaries between past, present and future are not inviolable barriers or borders built over time but part of a unified, holistic vision of the universe. We can begin to liberate ourselves from the narrow determinism imposed on our societies under monopoly capitalism, a cerebral straight-jacket designed to limit human possibilities instead of expanding them.

We need to resist the notion that our utopias can only be tangible and practical in the present so that we can convince pragmatists and skeptics that another city is really possible. The professional skeptics tend to worship neoclassical economic formulas, algorithms and probabilities but are absolutely ignorant of any language they were not taught. They are the leading ideologues of the Global North and cannot understand the dreams and real alternatives coming from the Global South. They are the professionals in engineering, design, architecture and planning who would lead us to more quick technological fixes. They are currently trying to sell us more electric vehicles instead of walkable cities, green golf carts instead of wild forests, recycling instead of waste reduction, and so many other greenwashing gimmicks. At the citywide level they propose sustainable cities that remain economically and socially unjust, resilient cities that exclude people of color and immigrants, green cities that are gentrified and exclusionary.

Quick fixes flourish in a global capitalist world that thrives on short-term thinking dictated by expectations of return on investment. Urban development obeys the real-estate investment cycle dictated by banks, real-estate investment trusts and individual investors. The planning horizon is at most several decades, not several centuries. We can learn from the many cultures and philosophies that do not make sharp distinctions between time and space and promote holistic, long-term thinking. For example, the Iriquois Nation in North America includes in its constitution a call to consider the next seven generations when making decisions.

In sum, we need to avoid painting one clear visual picture of the non-capitalist city because it is not a single vision or physical design we are after. Nor do we wish to substitute physical determinism for economic determinism. Nevertheless, if we are mindful of these pitfalls, imagining vivid images of different worlds can be powerful drivers of organizing at the grassroots. It is this kind of organizing across multiple spaces and nations that can generate the personal, political and structural changes needed to alter the course of human settlements.

To be more precise, the fundamental question is: how can our utopias help accelerate the struggles to transform and transcend the wasteful, inefficient metropolis built by capitalism? For example, our utopias can imagine replacing land developers and their teams of technocrats with *stewards* committed to protecting the earth and its inhabitants. Terms like "highest and best use of land" which under capitalist planning means the most profitable use, have to be retired and replaced by terms like *protecting Mother Earth* and new legal frameworks like *Earth Justice* (Cullinan 2011). The road to our utopias must also include proposals to reconstruct, deconstruct and eventually do away with the modern metropolis as we know it while establishing new social/spatial environments to meet the needs of humans, other species and the earth.

Our utopias must seek to repair the millennial damages done to people and the planet under colonialism, slavery and imperialism. This is not simply a challenge to change the physical city but a requirement that the city be reinvented and redesigned as a durable testimony to the crimes against humanity committed during the rise and fall of capitalism, colonialism and imperialism, and the resistance and bravery of the people who suffered and gave their lives in the struggles for economic, racial and environmental justice under capitalism. We must never forget the scope of the genocidal wars waged to secure mineral wealth, the millions who died in the mines, factories and workshops of powerful investors, or the half of humanity whose only option was to take care of homes and children so that capital could suppress wage levels and "keep women in their place." If we were to allow for the creation of new cities without acknowledging the bitter memories – including the atrocities, genocides and crimes against humanity and nature that were committed under the old regimes – the risk of receding into the dark past will remain. We only need to look at how in only half a century after the collapse of fascism in Europe revanchist ideas, political movements and politi-, cians considered to be "nationalists" and "populists" build on nostalgia and deep distress over the "loss" of their imperial metropolises and are reinventing fascism for the 21st century.

If capitalism's creative destruction began with the abolition of the countryside and rise of the city, a post-capitalist world will need to begin with a revolutionary new relationship between urban and rural settlements that is consistent with a revolutionary new economic and social structure in a post-capitalist world. As Marx and Engels pointed out in their path-breaking early works (Lefebvre 2016), the growth of industrial cities required the impoverishment of peasants who had been tied to rural food production so they could migrate to industrial cities and form part of the surplus labor pool that suppressed industrial wages. Marx noted how industrial fertilizers made peasant farming unfeasible and primitive sewage systems in cities created "waste" out of what otherwise would have continued to be natural fertilizers. Engels pointed out in *The Condition of the Working Class in England* (1973) and *The Housing Question* (1970) how the miserable housing conditions in the cities of capitalism were organically part of capitalist accumulation and would not be resolved simply by changing regulations, replacing landlords with universal homeownership, or passing narrow urban reforms.

The post-capitalist city has to reconnect urban environments with the natural world (Barlett 2005). It must reverse the destructive relationships that capitalism established to address the basic necessities of food, clothing and shelter, and its destruction of the natural environment. The metropolis of the future must produce most of its food instead of importing it from factory farms that use chemical fertilizers and pesticides that contaminate the land, and transport it long distances using more fossil fuels. Food should be a means for sustaining healthy life instead of a profit center for investors. Nations and people whose food systems were devastated by global monopolies must regain sovereignty over their food supplies. This will reduce the enormous transportation, environmental and public health costs resulting from the hegemony of global food monopolies. Basic necessities such as clothing should be supplied locally, reducing the vast inequalities that sustain global sweatshops. Would it be unreasonable that every household has a "rag-picker" able to recycle old clothing, a "gardener" to grow local food, and a trusty "handy-person" to maintain housing?

In the post-capitalist city, housing and land should cease to be commodities subject to the dictates of global finance. Housing should be a human right, along with health care and education. This will reduce the constant churning that leads to wasteful inefficiencies in the capitalist land and housing systems, including masses of construction waste, vacant buildings and sprawled metropolitan regions. It will put an end to the creative destruction of the capitalist city which leads to perpetual waste and redevelopment.

The future stewards of the metropolis must truly believe in and practice democratic planning at multiple scales that honors and respects the land instead of exploiting it for profitable investment. Investors and the real-estate industry will of course denounce this as an invitation for perpetual discussion and paralysis. The supply-side advocates who would solve the housing crisis by expanding the physical city and building more housing will ridicule it as irresponsible. But we know that the growth machine has only expanded shortages and inequalities while at the same time suppressing efforts to give a voice to people, communities and advocates for the earth.

Transforming cities into sustainable environments founded on equality and justice is going to be a thorough and *slow* process. Planning must be slow and continuing so that its many potential consequences can be evaluated. The non-capitalist city will be planned according to the *precautionary principle* – exploring potential long-term impacts of all human development and rejecting those strategies that harm the environment and reproduce injustices. Planning in the non-capitalist city will be *slow planning* because the pressures for short-term gains inherent in capitalism will no longer be there.

None of this can be sustained unless all people living in the city are guaranteed a right to the city and democratic participation in major decisions that affect the long-term future of the city and the planet. This requires the end of capitalism and reversal of the deeply harmful effects of colonialism and slavery, which have imbedded ethnic and racial oppression in the lifeblood of the modern metropolis. The exclusion of racial minorities from exercises in participatory planning is rooted not simply in exclusionary attitudes; it is a powerful legacy of structural racism and oppression (Angotti 2008).

In the post-capitalist city, access to education, health care, housing and a safe living environment must be guaranteed human rights, decreasing the need for market-driven wasteful consumption. Those who argue that it is impossible to provide high quality services to all unless there is dynamic capitalist growth that produces an abundant surplus have only to look at the performance of the United States among capitalist countries: it is far behind in the arenas of education, health care and housing when compared to other nations, especially other developed capitalist nations. It lags precisely because it is the global beacon of neoliberal capitalism.

Yes, but what will it *look* like?

I fear that all of my arguments will frustrate students and colleagues from professional fields such as my own – urban planning – who may very well agree with my structural and political arguments but end up hating me because I missed the opportunity to offer up some concrete, visible pictures that others can work with. As a brainy professor who has spent his lifetime moving between theory and practice, I can't dismiss this criticism because I deeply appreciate the value of having concrete visions and experiences that can help inform our utopias. Some of these will appeal to the architects, designers and planners who dwell on the physical city. While I ask them to avoid their occupational hazard of physical determinism (seeing changes to the physical city as necessary for changes to the social, economic and political city), I will indulge for a bit the prospect that, in fact, the physical city matters.

The post-capitalist city will minimize the effects of the human footprint on the land. Since land will no longer be a commodity, humans will need to reinvent ways they can relate to land without treating it as a thing. Humans will consider ways that they can be on and over the land by sharing land with other humans and the many other species that thrive on the land. In other words, land will have a use value and not an exchange value. There will be no space for gentrification and displacement. There will be green without green gentrification. This may result in more parks, open areas, paths for walking, but it may also result in more higher-density nodes for more intensive human activities.

The physical layout of the post-capitalist city will minimize the distances between home, work and the most widely used services. This will be possible when transportation systems no longer revolve around the use of personal vehicles. Electric vehicles won't do the trick; many studies and reports confirm that energy savings and greenhouse gas reductions resulting from electric vehicles will do very little to meet the goals of the UN climate accords (Sweeney & Treat 2019). The post-capitalist city will be a compact city, or a city of many compact districts. The big question, however, is whether the many districts will reproduce segregation and widen environmental injustices. The post-capitalist city must end the sharp geographic segregation into separate districts, ghettoes, Gold Coasts, gated communities and racial and ethnic enclaves. It will resolve the chronic problem of separation of workplace and residence by expanding mixed use districts, replacing factory towns with diverse local economies in which everyone willing and able to work can be healthy and productive.

The defenders of real estate's magical market logic often blame segregation and separation on deeply seated racial fears, especially in societies like those of Europe and North America that were born in colonialism, slavery and endemic racism. Or they will point their fingers at a few bad developers. However, it is really the structural imbeddedness of colonialism, racism and imperialism that normalizes urban segregation and makes it the unavoidable shaper of the modern metropolis.

Unlike the small towns that grew around mines and farming communities, the economic rationale for the post-capitalist city in the countryside will not be extractive capitalism. Mining will be highly regulated, avoiding destructive practices such as strip mining, the leveling of mountains and the discharge of wastes into local streams and aquifers. Food production will be more intensive, no longer confined to extensive factory farms producing only a small variety of products (soy, corn, wheat) using GMOs and chemical pesticides, and distances to markets and fuel costs will be minimized. In other words, we will eventually witness the elimination of the differences between city and countryside that have been the hallmark of capitalist development.

In reality, in everyday life the post-capitalist city will be indistinguishable in many ways from the post-capitalist countryside. The gross distinctions between the two that Marx and Engels identified will disappear.

Sprawl will disappear in the post-capitalist city because transportation systems will be designed to move the maximum number of people using the least amount of non-renewable energy. This means giving a priority to mass transit, bicycles, pedestrians and other forms of non-motorized transit such as scooters. It means more compact cities. Sprawl is a product of capitalism's conversion of land into a commodity; its urban land markets take the use value of centrality (a central location) and convert it into a market value. This drives many real-estate investors to the edges of cities where land is relatively cheap and where more homes can be built at relatively lower costs (relative to the central areas). The post-capitalist city will therefore find that efficient mass transit and non-motorized transportation are the most feasible alternatives. The inefficient private car will be used only when other alternatives are not feasible.

The post-capitalist city will reduce and eventually eliminate the antagonism between human settlements and nature. There will be no zoos. Domestic "pets" will no longer be caged and leashed. Around homes, humans will be the stewards of trees, flowers and shrubs and they will never use toxic chemicals and artificial fertilizers that contaminate the environment, only composted waste. In the post-capitalist city, a goodly portion of food needs will be met by local production and regenerative agriculture, reducing transportation costs and the use of fossil fuels.

The metropolis of the future will prioritize public transit, walking and biking, all of which are more energy efficient modes of transportation. It will be founded on ecological principles that follow the lead of indigenous people and environmental justice advocates (Bullard 2007). It will treat land and other species as cooperative partners and not commodities, use values and not exchange values. Cars will be used only for the transport of people with limited physical abilities. While mass transit allows residents to travel longer distances with greater speed, the rest of the transportation system will be more responsive to the need for *slow* travel.

The post-capitalist city will be a *slow city*. The transportation infrastructure and the entire city ought to be designed to facilitate the many human interactions that make sidewalks much more than thruways and turn them into *public places* where a wide array of human interactions spontaneously occurs. Workers will be free from speeded up production and consumption. Everyone will be able to enjoy a lengthy dinner every day with friends and family. Slowing down will abolish capital's historic drive for the 24/7 lifestyle and perpetual consumption.

As a result of the stunning technological advances of globalized capitalism, historic distinctions between time and space have been drastically reduced. However, far from reducing the need for travel it has stimulated more travel, made possible by the extremely low prices of fossil fuels. These prices are low because under capitalism the environmental and social costs of fossil-fuel production and consumption are "externalized," to use the term of neo-classical economics. We now know that the real long-term costs are significant and that their use may well have done permanent damage to the earth and its ecosystems and sealed the demise of the Anthropocene, taking with it the self-destructive mode of production that can no longer claim to be the beacon of global progress.

References

Angotti, T. 2008, *New York for sale: Community planning confronts global real estate*, MIT Press, Cambridge.
Angotti, T., 2012, *The New Century of the Metropolis: Urban enclaves and orientalism*, Routledge, NY.

Angotti, T., 2017, *Zoned Out: Race, Displacement and City Planning in New York City*. With Sylvia Morse UR Books, NY.

Angus, I., 2016, Facing the Anthropocene: Fossil Capitalism and the Crisis of the Earth System, Monthly Review Press, NY.

Baran, P. & Sweezy, P., 1966, *Monopoly Capital*, Monthly Review, NY.

Barlett, P.F. (ed.), 2005, *Urban Place: Reconnecting with the Natural World*, MIT Press, Cambridge.

Bellamy, E., 1960, *Looking Backward, 2000–1887*, American Library, NY.

Bonneuil, C. & Fressoz, J.-B., 2017, *The Shock of the Anthropocene*, Verso, London.

Bullard, R.D. (ed.), 2007, *Growing Smarter: Achieving Livable Communities, Environmental Justice, and Regional Equity*, MIT Press, Cambridge.

Caro, R., 1975, *The Power Broker: Robert Moses and the fall of New York*, Vintage, New York.

Cullinan, C., 2011, *Wild Law: A Manifesto for Earth Justice*, Chelsea Green, White River Junction, Vermont.

Engels, F., 1973, *The Condition of the Working Class in England*, Progress Publishers, Moscow.

Engels, F., 1970, The Housing Question, 1970, Progress Publishers, Moscow.

Foster, J.B., 2009, *The Ecological Revolution*, Monthly Review Press, NY.

Foster, J.B., Clark, B. & York, R., 2010, *The Ecological Rift: Capitalism's War on the Earth*, Monthly Review Press, NY.

Gould, K.A. & Lewis, T.L., 2017, *Green Gentrification: Urban Sustainability and the Struggle for Environmental Justice*, Routledge, Oxon, England.

Jacobs, J., 1961, *The Death and Life of Great American Cities*, Vintage, New York, NY.

Jacobs, J., 2004, *Dark Age Ahead*, Vintage, New York, NY.

Klein, N., 2014, *This changes everything*, Simon & Schuster, New York, NY.

Lefebvre, H., 2016, *Marxist thought and the city*, University of Minnesota Press, Minneapolis.

LeGuin, U., 1969, *The left hand of darkness*, Ace Books, New York, NY.

Marx, K. & Engels, F., 1947, *The German ideology*, International Publishers, New York, NY.

More, T., 2011, *Utopia*, W.W. Norton, New York, NY.

Morrison, T., 1987, *Beloved*, Knopf, New York, NY.

Roy, A., 2014, *Capitalism: A ghost story*, Haymarket Books, Chicago.

Strasser, S., 1999, *Waste and want: A social history of trash*, Henry Holt, New York, NY.

Sweeney, S. & Treat, J., 2019, *The road less travelled: Reclaiming public transport for climate-ready mobility*, Trade Unions for Energy Democracy, Working Paper No. 12. Trade Unions for Energy Democracy (TUED), Rosa Luxemburg, New York.

Sze, J., 2007, *Noxious New York: The Racial Politics of Urban Health and Environmental Justice*, MIT Press, Cambridge.

10 Towards democratic and ecological cities

Yavor Tarinski

Democratic institutions

The basis on which such democratic and ecological cities should be built is a process of constant popular self-institution. In other words, this implies the collective creation of participatory decision-making bodies, through which the citizens to be able directly to shape the laws and rules of their common urban habitat. Such democratic institutions will be nothing like the current bureaucratic ones that keep power away from the grassroots, but they also go beyond the anarchist slogan "make war on institutions, not on people",[1] which implies that institutions as such are the obstacle to emancipation and self-determination. An institutional structure, based on direct citizen participation, will allow people to self-limit their activities so as for genuine freedom to emerge. This comes in line with Rousseau's idea that *the impulse of mere appetite is slavery, while obedience to a self-prescribed law is liberty*.[2]

People from the independent cities of the Antiquity and the Middle Ages sought, through such self-institution of laws and constitutions, protection from kings, tyrants, nobles and oppressors. The oligarchic structure of the modern city, on the other hand, allows for growing inequality and precarity to reign. This is so due to the fact that the power is being centralized in the hands of the bureaucratic and business elites, while the citizens' participation is being limited to vote-casting in elections.

The role of such democratic institutions is to make the exercise of nonstatist, nonoppresive and noncapitalist power possible. Suitable decision-making bodies for this framework are the neighborhood assemblies and municipal councils, as demonstrated by the ancient Athens. The ancient Athenians based their city management on the *ekklesia* – popular assembly in which all citizens had the right to directly participate in the management of public Affairs – while choosing their magistrates by the means of sortition (choosing by lot), in order to avoid demagoguery or professionalization of the political realm. While the Athenian society had many shortcomings – like slavery and patriarchy – it still offered us the concept of the polis: a free city managed directly by its citizens.

The political foundations of our cities then, could be based not on centralized bureaucratic mechanism, but on network of popular assemblies, each one operating in neighborhoods or areas with population between 30,000 and 50,000 (the amount of citizens in ancient Athens that had the right to participate in the general assembly). These bodies will be the main locus of power, through which the citizens will shape the common framework of policies and laws to which all urban dwellers should abide to.

Besides the assemblies, there is the need of municipal councils as a supplement for the exercise of grassroots power. Their members could be chosen by lot, following the democratic tradition, and remain revocable at any time if deemed that they exploit their position. Such institution will deal with routine tasks and will be responsible for monitoring the implementation of the decisions, taken by the general assembly. The councils should hold their meetings (which must be public) as often as necessary (for example twice a week). The regular rotation of delegates (once every two, three months or more) will prevent the emergence of a hierarchy and will allow broader participation in the council.

One such democratic and ecological project is genuinely stateless, and thus strives to connect cities by means that radically differ from those of the nation-state. Instead, it unites them in confederations, in which every city maintains its sovereignty. Murray Bookchin describes that one such democratic confederation

> should be regarded as a binding agreement, not one that can be canceled for frivolous "voluntaristic" reasons. A municipality should be able to withdraw from a confederation only after every citizen of the confederation has had the opportunity to thoroughly explore the municipality's grievances and to decide by a majority vote of the entire confederation that it can withdraw without undermining the entire confederation itself.[3]

Post-capitalist municipalized economies

The paradigm of democratic and ecological cities implies that the economy will be municipalized, i.e. directly owned and managed by the citizens. In one such city, economic activities will be placed in the hands of the urban communities, under the direct control of the popular assemblies, councils and confederations. As author and activist Janet Biehl suggests, in this way *the citizens would become the collective "owners" of their community's economic resources.*[4]

In one such paradigm, the inhabitants of the city don't vote as workers and/or consumers. Instead they participate directly, as citizens, in the formulation and approval of economic policies regarding their neighborhoods and city. The citizen body will collectively determine its needs, as well as distribute the material means of life, decide how to use available recourses etc. The democratic institutions will allow for everyone in the community to have access to the means of life, regardless of the work he or she was capable of performing. Furthermore, with citizens forming collectively the economic policies of their city, there will be no space left for capitalist antagonism, as all economic entities will have to adhere to ethical percepts of cooperation and solidarity.

Regarding the economies of wider regions, which include more than one city, it will be up to the democratic confederations, as described earlier, to exercise power "from below". As Biehl writes[5]:

> the wealth expropriated from the property-owning classes would be redistributed not only within a municipality but among all the municipalities in a region. If one municipality tried to engross itself at the expense of others, its confederates would have the right to prevent it from doing so. A thorough politicization of the economy would thereby extend the moral economy to a broad regional scale.

The municipalized economy of the post-capitalist city ceases to be, as Bookchin suggests,[6] an economy in the strict sense of the word. Instead, it becomes incorporated within the direct democratic political processes of the community, as it is democratically guided by humane and ecological standards. In one such economy, where people participate primarily as citizens, an ethos of public responsibility emerges. Since it is not only economic prosperity, in the narrow sense of the term, which is being sought, a more general quality of urban life, which includes things like healthy environment and strong communal bonds, becomes the prime target. As Andrew Flood wrote in 1995,[7] *in a society where we democratically control production, we will decide not to pollute, or to limit pollution to a level that can be absorbed.* His suggestion stems from a post-growth logic, according to which the increase in production and consumption is not an end in itself. In such democratic and ecological paradigm, stewardship and self-limitation, aiming at the general increase of quality of life, replace the capitalist imperatives of unlimited expansion and competition for short-term profit.

Agriculture

The relationship between agriculture and the city will be radically transformed in the transition towards post-capitalist democratic and ecological future. Nowadays food production is zoned away from urban areas. The food which reaches our cities arrives from distant areas in huge containers, produced by multinational industries, which only care for their narrow profits. There is much insistence on the way food is being packaged and promoted rather than its quality. Even products that are labeled biologically or organically produced are intended to make money, rather than increase the health of as much people as possible.

Nowadays those who run these agricultural industries have little contact or knowledge with/about the land on which their employees are cultivating. In fact, the way this domain is being managed does not differ considerably than any other wasteful, short-term oriented capitalist industry. This comes in stark contrast with earlier form of agriculture. Ancient cities formed around farmlands. Their people viewed food cultivation as a spiritual activity, and its consumption, as social ritual. There was this attitude of stewardship, rather than exploitation. The people of ancient Athens, like Theophrasus, regarded the interaction of society with nature as relationship between two autonomous equal entities.[8] For indigenous people like the Cayuses in North America, the ground beneath them was alive, and they listened to it, in order to hear the "Great Spirit".[9]

The high esteem the earth had in many of these societies had to do with their dominant paradigm. Many of them were rather feminine, in the sense that feminist conceptions of symbiosis within society and with nature were the basis of their worldview (except of the Greeks for example, but their departure towards patriarchy and gerontocracy was not as temporally expanded as it is for our modern societies). As it was the ground from which they cultivated their food, they considered it the mother of all life. Even today indigenous communities like the Zapatistas, which live close to their land, have built their mythology around the earth (and corn especially, as one of their main crops).

In one democratic and ecological future, where city life does not equal environmental degradation, agriculture is being integrated to a certain degree within the urban matter, while non-urban agricultural areas, which will still be vital for the feeding of the citizenry, will have to be integrated into democratic confederations, alongside self-managed cities, which means radical decentralization of the agricultural production into cooperatives and rural assemblies. This implies new democratized approach to agriculture, which as Bookchin suggests:

> *transcends the prevailing instrumentalist approach that views food cultivation merely as a "human technique" opposed to "natural resources." This radical approach is literally ecological, in the strict sense that the land is viewed as an oikos – a home. Land is neither a "resource" nor a "tool," but the oikos of myriad kinds of bacteria, fungi, insects, earthworms, and small mammals. If hunting leaves this oikos essentially undisturbed, agriculture by contrast affects it profoundly and makes humanity an integral part of it. Human beings no longer indirectly affect the soil; they intervene into its food webs and biogeochemical cycles directly and immediately.*[10]

Jane Jacobs explains one such relationship between the urban and the rural in the following symbiotic way: *Big cities and countryside can get along well together. Big cities need real countryside close by. And countryside – from man's point of view – needs big cities, with all their diverse opportunities and productivity, so human beings can be in a position to appreciate the rest of the natural world instead of to curse it.*[11]

In the paradigm of democratic and ecological cities, nature is interwoven within the urban matter. This means that it will be present into the everyday life in an essential manner, unlike today, where large parks are among the few interactions a person can have, and not on a daily basis. While there will be need of certain urban planning towards such natural integration, it will be mostly up to the democratized social, political and economic relations. For example collectives and individuals could be producing food on unused urban surfaces like terraces, rooftops, parking lots, etc. This would also imply that networks of free, pollution-free, public transport should be expanded to such an extent as to liberate significantly the streets from car traffic (and consequently from the need for parking lots).

The integration of agriculture into the urban life will increase the food sovereignty of cities. But surely it will be not enough. Cities will have to establish a new type of relationship with the countryside (where most agriculture is taking place), based on collaboration and mutual aid, instead of domination and profiteering.

Such relations will have to be based on democratic confederations which allow to all involved to maintain their political sovereignty. This means that villages and rural towns will have to adopt the direct democratic approach of the democratic and ecological cities. Through this confederal level urban dwellers could engage in participatory planning, regarding their needs, and send them to their rural allies and vice versa.

Energy

The question of energy is of crucial importance when we discuss the future of our cities. Our contemporary heavily urbanized societies consume huge amounts of energy, which is being derived through environmentally degradative means.

The creation of democratic and ecological cities requires departure from our current energetic paradigm. Instead it implies changes in two basic directions: first, by going beyond the logic of technological neutrality, and second, by rethinking how our needs are being formed and towards what ends.

For the modernist Left, the problem is not our current technology but who owns it. For them technology is violent, wasteful and destructive only when used by the wrong hands (for example those of the capitalists). In their view, every technological innovation is not shaped by the context in which it was created, which in itself is really problematic view.

Driven by this logic, many on the Left imagine the post-capitalist city's energetic needs being supplied by nuclear power. Their answer to the anti-nuclear movement is that our current dependence on fossil fuels is destroying our world and nuclear energy is the quickest way towards salvation. But they seem blind to the characteristics, this energy source has, which were embedded in it by the contextual environment in which it emerged.

First of all, nuclear power is incompatible with decentralized and democratic forms of self-governance. Instead, as suggested by researcher Aaron Vansintjan, it requires large state subsidies and centralized planning.[12] According to him:

> *Nuclear power requires a regime of experts to manage, maintain, and decommission; a centralized power grid; large states to fund and secure them; and, then, a stable political environment to keep the waste safe for at least the next 10,000 years. The technology is only 80 years old, modern states have existed for about 200, humans have only been farming for 5,000, and most nuclear waste storage plans operate at a 100-year time-span. To put it mildly, an energy grid dependent on nuclear means having lot of trust in today's political institutions.*

This is deeply political issue. The vision of a nuclear-powered society implies the creation of a totalitarian-like organizational structure, a powerful state. The scale of such an energy system demands to be situated away from the people, in areas zoned away from the rest of society (even whole cities build around such power plants). In this environment scientists and technocratic elites will naturally play an important role. With all the dangers that come with nuclear power plants, there will be need of high level security measures, control and supervision. All these requirements make nuclear energy incompatible with direct democratic ecological visions. Instead, it is much more suitable for totalitarian ones like eco-fascism.

Furthermore, nuclear energy is incompatible with the new climate-impacted planetary conditions, which are highly prone to fires, extreme storms and sea-level rise. With the increase of the probability of environmental catastrophes and extremities, it is questionable to say the least, whether nuclear power can function safely. Professor Heidi Hutner has pointed out that wild weather, fires, rising sea levels, earthquakes and warming water temperatures all increase the risk of nuclear accidents.[13] And on top of that, the lack of safe, long-term storage for radioactive waste remains a persistent danger.

An energy source, compatible with the paradigm of democratic and ecological cities is the one derived from renewables. But simply shifting from fossil fuels to renewable sources will not suffice. We must, first of all, avoid approaching renewables from a modernist perspective. This would mean that we cannot use them mainly in a centralized manner (like industrial-style enormous solar or wind farms), since this would require a bureaucratic managerial apparatus, not much different from the one required by nuclear power. Although it will never be possible to avoid larger scales, one democratic and ecological paradigm would require for us to develop renewables towards the greatest possible decentralization, so as to allow local communities to have direct control over their energy supply.

Then there is another issue that must be seriously considered. As author Stan Cox notes:

> *There's nothing wrong with the '100-percent renewable' part . . . it's with the '100 percent of demand' assumption that [scientists] go dangerously off the rails. At least in affluent countries, the challenge is not only to shift the source of our energy but to transform society so that it operates on far less end-use energy while assuring sufficiency for all. That would bring a 100-percent-renewable energy system within closer reach and avoid the outrageous technological feats and gambles required by high-energy dogma. It would also have the advantage of being possible.*[14]

Thus from ecological and democratic perspective, we cannot simply switch to this or that technology. We have to bear in mind the contextuality of every technological innovation and the scale on which it is being implemented. Energy is much more than simply a tool: it has to do with relationships between people, societies and ultimately between humanity and nature. Furthermore, it is not just a means for the satisfaction of our needs, but a need in itself, and in a democratic paradigm, it will have to be deliberated on grassroots level by all members of society.

Democratic and ecological urban design

The creation of democratic and ecological cities is a complex thing. For society's organization to be reorganized on the basis of direct democracy and environmental sustainability, among the many preconditions that seems to be required, is the breaking of alienation and establishment of communalist relationships within the urban realm. A city that would encourage and strengthen community feeling would represent mixture of housing, public, workplace, shopping, green and other spaces, all of which will be within walking distance or reachable by public transportation, in contrast with the modern mainstream way of urban design, based on positioning of fixed zones across vast distances.

A mixed architecture consisted of medium-sized housing cooperations, with adjoined gardens, within a walking distance from schools, public squares, markets and green spaces will allow for the experience of random interactions between neighbors. The walking element could build feeling of belonging to the city, with citizens developing strong links with their local, social and urban environment. It will also, as author Jay Walljasper suggests,[15] contribute for greater economic equality by allowing everyone the right to freely move across the city, without the need of car.

The shift towards walkable cities would imply the radical rethinking and remaking of roads and streets, today designed mainly as high-speed arteries connecting housing districts with office areas, encouraging driving over walking. As Donald Appleyard's famous 1972 study demonstrates,[16] as heavier the car traffic on one street is, as less are the walkers and the everyday communal experiences. This, except the obvious effects on human health (leading to obesity, heart diseases, etc.), contributes to the already high levels of alienation in urban areas.

An approach that could alter this alienating effect, encouraging instead people to walk on the streets and potentially to produce community feeling is the narrowing of streets in urban areas, expansion of pedestrian spaces, introduction of wider bicycle alleys etc. As the city planner and author Jeff Speck explains[17] *people drive faster when they have less fear of veering off track, so wider lanes invite higher speeds*. This, in mixture with vast network of free urban public transportation, will allow for daily social interactions on them by pedestrians and passengers. The daily social experiences like nodings, smiles and random chatting with co-citizens potentially could contribute for us to feel more comfortable on our streets.

This would bring with itself other positive features, like drastic reduction of the health problems mentioned earlier, but also with reduction of car speed, responsible for the death of huge number of people around the world, as well as reduction of air pollution of the contemporary private car dominated metropolises.

Green spaces are another key aspect of the urban environment. According to Bob Lalasz,[18] they tend to make people happier. Furthermore, green spaces bring people closer together. Thus in one democratic and ecological urban project nature should be essential part of the urban landscape. The gardens, part of housing cooperations will allow for the experience of gardening time by neighbors, bonding them together. It will also potentially encourage the development of communal/solidarity economy, by producing their own food and exchanging it or sharing it with other urban gardeners.

But except them, parks and public gardens should be shuffled across the mixed urban architecture. There is a certain trend in the modern metropolitan cities for large-scale parks to be zoned away from housing districts and office areas, making human interaction with nature a rare opportunity. Contrary to this logic, the mixed city, described here, could propose green spaces located in various locations across the city. As Charles Montgomery suggests,[19] this does not exclude the existence of large-scale parks, but that the urban green space will not be limited to them. This will imply that people will have the possibility to get in contact with tiny gardens and parks on their way, let's say, to work etc., as well as experiencing the feeling of being "into the wild" by entering the huge local parks.

The squares play a key role in one city that encourages democratic and ecological culture, since they act as spaces for social interactions as well as forums for expression of civic opinions. Thus they should be made freely available for popular interventions, unlike today, where bureaucrats decide who, when and for what reason should organize an event in them.

But we also hear critiques about overcrowdedness of modern cities, leading to further alienation and withdrawal into passivity. If this is true, should we abandon city life altogether and return to village life? According to psychologist Andrew Baum's study,[20] the feeling of overcrowdedness is being fed by design that does not allow people to control the intensity of spontaneous social interactions. Baum compared the behavior of residents of two very different college dormitories. He concluded that students whose environment was allowing them to control their social interactions experienced less stress and built more friendships than students who lived along long and crowded corridors.

Therefore an answer to the "overcrowdedness" problem could be found in the creation of semi-public/communal spaces, which to represent a middle passage between the private and the public. This would imply the abandonment of the gigantic housing projects in which large numbers of people live together (like the socialist-era gigantic worker "barracks"), never feeling quite alone. Instead, a space could be given to medium-size housing cooperations, with common spaces, in disposal of all the neighbors. Thus three layers of social spheres will be creating – private, communal and public – allowing citizens to regulate their social interaction, thus giving them sense of comfort and encouraging egalitarianism.

Urban design and direct democracy

Of course, many things can be done with urban design to encourage communal feeling across citizens, but it cannot alone do this job without providing space for institutions of public deliberation, which to enable for co-inhabitants collectively to determine the destiny of their cities as well as of themselves. It is difficult to imagine what else could bring people more close as community than the feeling of shared responsibility for their city.

Thus a democratic and ecological city should always strive at managing itself through direct democracy. This will require the establishment of public spaces, suitable for the accommodation of direct-democratic institutions, like the popular assemblies described earlier. Such spaces, like public squares, halls or amphitheatres, should most likely, be equipped with sound systems, allowing for single speaker to be heard amongst gatherings of several thousand citizens, as well as live streamed for the rest of the community to be able to observe from distance.

Murray Bookchin points[21] at the cities of the past, before the emergence of the so called statecraft. In them the citizens were actively involved in shaping their cities, deeply and morally committed to them. But with the emergence of parliamentarism and capitalism, they were replaced by passive consumers, simply passing through their urban environment, without any commitment to it.

Such step towards reframing the city's role as encourager of community and citizenry is, in a sense, rediscovering the ancient Athenian logic of the *polis*. Of course the sizes of their times and our own are incomparable, but the logic on which their city was build could be used as "germ", as suggested by Cornelius Castoriadis,[22] by us today. Ancient Athens was encouraging community feeling as well as active citizenry, which gave birth to one of the most influential periods of human creativity to this day. At the heart of the Athenian urban life were situated the agora and the general assembly. The agora was a market place, positioned in accessible and central part of the city, where the Athenians spent great deal of their time exchanging goods, information and opinions, or in other words – socializing – while in the assembly they were bonding with each other as well as with their city by sharing responsibility for its destiny.

Towards a strategy from below

The paradigm of democratic and ecological cities is not just a utopian vision for a future never to come. There are countless grassroots initiatives and struggles that strive towards that goal from today. During the last years we are even witnessing a rising interest among social movements in the urban question. More and more people are starting to notice the effects our cities have on us. Different movements, focused on the urban question, are emerging, some focused on municipal elections, others on urban planning. But it seems that most of them do not view this matter in holistic political manner.

On the one hand, the introduction of changes, no matter how great, in the way local elections are being held, won't give cities back to their citizens. This can be done only by introducing new deliberative institutions, such as the described earlier in this text, which will allow to each and every citizen directly participating in the determination of his city's destiny. The role of existing local authorities should be reduced to supervision and enforcement of the decisions, already taken by these new institutions, and thus subjected to them through means such as revocability, sortition, rotation, etc.

On the other hand, often social movements dealing with city issues tend to limit their activities to narrow urban designing, waiting from local authorities to implement their proposals. Their work remains half-way done, since a city is not consisted only of buildings, roads and squares, but also of people and thus, social relations and forms of organization. As Henri Lefebvre suggests[23]: *The right to the city cannot be conceived of as simple visiting right or as a return to traditional cities. It can only be formulated as a transformed and renewed right to urban life.*

Thus the approach should be focused on linking urban design with politics and decision-making in particular. As we saw earlier, radical change in the one is difficultly imaginable without such radical change to occur and in the other. But what seems a very good start is the fact that more people are paying attention to the role our urban environment is playing on us, our social relationships and our political projects in general.

Conclusion

The creation of democratic and ecological cities is a question of radical social transformation. The city has played an important role in human life from antiquity until nowadays. We cannot think of our future without thinking of the future of our urban inhabitant. Oversimplified proposals for the abandoning of city life and retreat to small villages and rural life have either lost touch with reality, or are being influenced by primitivism and its anti-political orientation.

If we are to create democratic and ecological society, we will have to rethink and remake our cities along democratic and ecological lines. We must depart from the elements which have negatively shaped the modern city: modernist thinking (large scales and centralization), unsustainable and short-term profiteering (fossil fuels and monocultures), and exploitation (vertical management and capitalism). New principles must be adopted, such as environmental sustainability (citizens acting as conscious stewards) and democratic participation (all citizens shaping directly and collectively their common city life).

Ultimately, the question of democratic and ecological cities is a deeply political one, as it requires the collective deliberation of the future we want. This includes not only deciding how we would like our common world to look like, but also what characteristics we wouldn't like to see in it. And this ultimately is a question of self-limitation – something impossible within the framework of capitalism and statecraft. As Aaron Vansintjan concludes,[24] talking about limits isn't constraining, it's liberating – perhaps paradoxically, it's the basic requirement for building an ecological future of real abundance.

Notes

1 Bookchin, M. (1999). *Thoughts on libertarian municipalism* [Online], Institute for Social Ecology. https://social-ecology.org/wp/1999/08/thoughts-on-libertarian-munic-ipalism/ (Accessed: 04 October 2020).

2 Rousseau, J. J. (1998). *The Social Contract*. Ware: Wordsworth Editions. p. 20.

3 Bookchin, M. (1999). *Thoughts on Libertarian Municipalism* [Online]. Institute for Social Ecology. Available at: https://social-ecology.org/wp/1999/08/thoughts-on-liber-tarian-municipalism/ (Accessed: 4 October 2020).

4   Biehl, J. (1998). *The Politics of Social Ecology: Libertarian Municipalism*. Montreal: Black Rose Books. p. 118.

5   Biehl, J. (1998). The Politics of Social Ecology: Libertarian Municipalism. Montreal: Black Rose Books. p. 120.

6   Bookchin, M. (1986). *Municipalization: Community Ownership of the Economy* [Online]. Anarchy Archives. Available at: http://dwardmac.pitzer.edu/Anarchist_Archives/bookchin/gp/perspectives2.html (Accessed: 4 October 2020).

7   Flood, A. (1995). *Anarchism and the Environmental Movement* [Online]. The Struggle Site. Available at: http://struggle.ws/talks/envir_anarchism.html (Accessed: 4 October 2020).

8   Hughes, J. D. (1975). 'Ecology in Ancient Greece', *Inquiry: An Interdisciplinary Journal of Philosophy (18/2)*. pp. 115–125.

9   Debo, A. (1984). *A History of the Indians of the United States*. Norman: University of Oklahoma Press. p. 157.

10   Bookchin, M. (2014). *Radical Agriculture* [Online]. Libcom. Available at: https://libcom.org/library/radical-agriculture-murray-bookchin (Accessed: 04 October 2020).

11   Jacobs, J. (1992). *The Death and Life of Great American Cities*. New York: Vintage Books. p. 447

12   Vansintjan, A. (2018). *Where's the "Eco" in Ecomodernism?* [Online]. Red Pepper. Available at: www.redpepper.org.uk/wheres-the-eco-in-ecomodernism/ (Accessed: 4 October 2020).

13   Hutner, H. (2019). *Nuclear Power is not the Answer in a Time of Climate Change* [Online]. Aeon Magazine. Available at: https://aeon.co/ideas/nuclear-power-is-not-the-answer-in-a-time-of-climate-change (Accessed: 4 October 2020).

14   Cox, S. (2017). *Renewables Alone Won't Save Us* [Online]. Dissent Magazine. Available at: https://www.dissentmagazine.org/online_articles/100-percent-renewable-energy-overconsumption-inequality (Accessed: 4 October 2020).

15   Walljasper, A. (2015). *A Good Place for Everyone to Walk* [Online]. Common Dreams. Available at: www.commondreams.org/views/2015/10/23/good-place-everyone-walk (Accessed: 4 October 2020).

16   Appleyard, D. and Lintell, M. (1972) 'The Environmental Quality of Streets: the Resident's View Point', *Journal of the American Planning Association*. pp. 84–101

17   Speck, J. (2014). *Why 12-Foot Traffic Lanes Are Disastrous for Safety and Must be Replaced Now* [Online]. Bloomberg CityLab. Available at: www.citylab.com/design/2014/10/why-12-foot-traffic-lanes-are-disastrous-for-safety-and-must-be-replaced-now/381117/ (Accessed: 4 October 2020).

18   Lalasz, B. (2015). *Go to Your Happy Place: Understanding Why Nature Makes Us Feel Better* [Online]. Cool Green Science. Available at: http://blog.nature.org/science/2015/05/22/science-nature-emotion-affect-feel-better/ (Accessed: 4 October 2020).

19   Montgomery, C. (2015) *Happy City: Transforming our Lives Through Urban Design*. London: Penguin Books. p. 110

20   Valins, S. and Baum, A. (1973) 'Residential Group Size, Social Interaction, and Crowding', *Environment and Behavior*.

21   Bookchin, M. (2014) *Toward a Communalist Approach* [Online]. New Compass. Available at: http://new-compass.net/articles/toward-communalist-approach (Accessed: 4 October 2020).

22   Curtis, D. A. (eds.) (1997) *The Castoriadis Reader*. Oxford: Blackwell. p. 269

23   Lefebvre, H. (1996) *Writings on Cities*. Oxford: Blackwell. p. 158

24   Vansintjan, A. (2018). *Where's the "Eco" in Ecomodernism?* [Online]. Red Pepper. Available at: www.redpepper.org.uk/wheres-the-eco-in-ecomodernism/ (Accessed: 4 October 2020).

References

Appleyard, D. & Lintell, M., 1972, 'The environmental quality of streets: The resident's view point', *Journal of the American Planning Association*, 84–101.

Biehl, J., 1998, *The politics of social ecology: Libertarian municipalism*, Black Rose Books, Montreal.

Bookchin, M., 1986, *Municipalization: Community Ownership of the Economy* [Online]. Anarchy Archives, viewed 4 October 2020, fromhttp://dwardmac.pitzer.edu/Anarchist_Archives/bookchin/gp/perspectives2.html.

Bookchin, M., 1999, *Thoughts on Libertarian Municipalism*, Institute for Social Ecology, viewed 4 October 2020, fromhttps://social-ecology.org/wp/1999/08/thoughts-on-libertarian-municipalism/.

Bookchin, M., 2014a, 'Toward a communalist approach', *New Compass*, viewed 4 October 2020, fromhttp://new-compass.net/articles/toward-communalist-approach.

Bookchin, M., 2014b, 'Radical agriculture', *Libcom*, viewed 4 October 2020, fromhttps://libcom.org/library/radical-agriculture-murray-bookchin.

Curtis, D.A. (eds.), 1997, *The Castoriadis Reader*, Blackwell, Oxford.

Debo, A., 1984, *A history of the Indians of the United States*, University of Oklahoma Press, Norman.

Flood, A., 1995, 'Anarchism and the environmental movement', *The Struggle Site*, viewed 4 October 2020, fromhttp://struggle.ws/talks/envir_anarchism.html.

Hughes, J.D., 1975, 'Ecology in ancient Greece', *Inquiry: An Interdisciplinary Journal of Philosophy* 18(2), 115–125.

Hutner, H., 2019, 'Nuclear power is not the answer in a time of climate change', *Aeon Magazine*, viewed 4 October 2020, fromhttps://aeon.co/ideas/nuclear-power-is-not-the-answer-in-a-time-of-climate-change.

Jacobs, J., 1992, *The death and life of great American cities*, Vintage Books, New York, NY.

Lalasz, B., 2015, *Go to Your Happy Place: Understanding Why Nature Makes Us Feel Better* [Online], Cool Green Science, viewed 4 October 2020, fromhttp://blog.nature.org/science/2015/05/22/science-nature-emotion-affect-feel-better/.

Lefebvre, H., 1996, *Writings on Cities*, Blackwell, Oxford.

Montgomery, C., 2015, *Happy city: Transforming our lives through urban design*, Penguin Books, London.

Rousseau, J.J., 1998, *The social contract*, Wordsworth Editions, Ware.

Speck, J., 2014, *Why 12-foot traffic lanes are disastrous for safety and must be replaced now*, Bloomberg CityLab, viewed 4 October 2020, fromwww.citylab.com/design/2014/10/why-12-foot-traffic-lanes-are-disastrous-for-safety-and-must-be-replaced-now/381117/.

Valins, S. & Baum, A., 1973, 'Residential group size, social interaction, and crowding', *Environment and Behavior* 5(4), 421–439.

Vansintjan, A., 2018, 'Where's the "eco" in ecomodernism?', *Red Pepper*, viewed 4 October 2020, from www.redpepper.org.uk/wheres-the-eco-in-ecomodernism/.

Walljasper, A., 2015, *A good place for everyone to walk*, Common Dreams, viewed 4 October 2020, from www.commondreams.org/views/2015/10/23/good-place-everyone-walk.

11 The coming revolution of peer production and the synthetisation of the urban and rural

The solution of the contradiction between city and the country

Jakob Rigi

Introduction

The formation of mega cities, their wealth and power and the flows between them seem to dominate the global spatial dynamics of the contemporary capitalism (Castells 2010). This chapter argues that a new non-capitalist mode of production namely the Commons-Based Peer Production (CBPP) is emerging (see below). This mode of production can reverse the spatial logic of the mega city. Its global space will consist of urban–rural interconnected localities whose population will not need exceed 500,000 individuals. These localities are centres of culture, innovation and creativity, cosmopolitan and globally connected. They also have very strong rural features. They preserve and synthesise the good sides of city and the country and negate their negative aspects.

The rest of this article: 1) provides a historical perspective on built space; 2) describes the current embryonic form of CBPP; 3) speculatively describes the contours and features of a new global civilisation that will be based on a fully fledged CBPP; 4) describes the principle of the spatial dynamics of this civilisation as a combination of ruralisation-urbanisation processes; and 5) in the conclusion it comments on and defend the viability of CBPP and its corresponding spatial project.

Historical perspective

Hunter and gathers had a highly sophisticated practical-spatial orientation of their natural habitat. They marked and named residential, hunting and gathering areas, different places in forest, hills, springs and revivers, etc. (Turnbull 1961).

The production of surplus by the direct producers resulted in the division between manual and mental labour emerging on the basis of the gender division of labour (Engels 1968). Since, this surplus permitted some to engage in political and intellectual activities without toiling to produce their own livelihood (Aristotle 1885, 1996). This led to the formation of classes, private property and state (Engels 2000; Smith 1984).

The peasant village with it houses and farms and surrounding pastures, sources of water, woods, fisheries, etc. more or less constituted a permanent and relatively enclosed locality corresponding to the community of the villagers.

The classical political city (Athens and Rome) were administrative centres of power fed by the surplus produced in the countryside. The same was true of the oriental city (for instance, the medieval Baghdad) (Lefebvre 2003). These cities also included crafts and commerce, though in the European city-states [Athens and Rome] merchants and artisans had low social statuses (Lefebvre 2003).

It was in the late medieval Europe that city emerged as especial site of commerce and craft antagonistically opposed to feudal fort. Serfs fleeing the feudal repression in the countryside sought refuge in the city. They constituted an important portion of its population. The mercantile city of 16th and 17th centuries, politically and to some extent economically dominated the countryside (Lefebvre 2003). However, economically it was dependent on a surplus that was overwhelmingly produced in the countryside in the framework of pre-capitalist modes of production. This surplus was extracted the combination of trade with the exploitation of indented labour, murder and plunder on an international scale (Woolf 2010; Wood 2003). Thus, this city was the centre of a loosely integrated global spatial order whose periphery consisted of the "national" and international areas from which the surplus was extracted.

The industrial capitalism introduced an epoch-making spatial revolution. It created a solidly integrated and dynamic, though uneven and contradictory, global spatial order. This order was geared to the production of surplus-value and its realisation and distribution between different strata of exploiting classes (Marx 1976; Lefebvre 2003; Smith 1984). Neil Smith (1984:136–148) argues that this order, according to Smith still prevailing, is constituted through dialectical relations between the three "urban", "global" and "nation-state" "scales". The urban scale is the site of the production and reproduction of labour force (ibid: 137), and the global scale is generated by the quest of capital for the global levelling of labour power and "universalisation of wage labour relation" (ibid: 139). The nation-state scale results from the competition between different nationally based capitals "in the world market" (ibid: 142). Harvey (2003) introduces the regional scale. He emphasises the centrality of regional scale for the capitalist spatial order: certain regions due the concentration of infrastructure, skilled workers and institutional arrangements, etc. provide favourable conditions for the accumulation of capital for certain periods of time, though they may lose this significance.

Manuel Castells (2010) argues that the IT revolution has led to the emergence of space of flows consisting of a network of "global cities" (Sassen 1999) interconnected through the follows of knowledge, capital, goods and people. These cities are site of the production of advanced knowledge/information, manufacturing, real estate, financial and insurance services and creative and cultural industries. This global space of flows subjugates and marginalises the experience of locality. This spatial antagonism, the space of flows and locality corresponds to, constituted by and constitutes a social antagonism between the globally cosmopolitan elites and the marginalised social groups deprived of fruits of "the Network Society". Technological infrastructure of electronic communication, financial institutions and managerial and creative elites are the pillars of the space of flows (442–43, 455).

The industrial city spurred the dramatic increase in the urban population and expansion of the urban space. These demographic and geographic tendencies have peaked in the era of the global city by the formation mega urbanities especially in China (Castells 2010).

Uneven development has been the core principle the capitalist spatial dynamics (Mandel 1975; Harvey 1982; Smith 1984). First, there is a division in term of wealth, infrastructure, financial and social resources between the global North and the global south. Second, there is an unevenness between rich urban centres and small towns and rural areas in each country. And, third, there is a huge polarisation between wealth and poverty within global cities (Harvey 2012). Generally, the elite centres of the global cities subjugate and exploit the rest of the world. Extraction of rent – including knowledge rent and advertising rent – is a major mechanism of this exploitation (Harvey 2012; Rigi 2014a; Rigi & Prey 2015). Thus, the global cities are explosive contradictions. On the one hand these cities are the centres of progressive social movements, creativity, invention and artistic development, and on the other hand sites of vanity, conspicuous consumption, spiritual and material poverty, crime, pollution, exploitation and alienation (Harvey 2012).

This chapter argues capitalism and the mega city underpinned by the capitalist quest for making profit are historically superfluous. They are ridden with crises and antagonism. Thus, the desire of environmental and other social movements to reconstruct the logic of space around democratic localities is in tune with paradigmatic productive forces of our era, namely IT and science. The capitalist relation of production prevents the optimal development of science on the one hand and destroys nature which is the most elementary productive force on the other. The knowledge based forces of production have spurred the emergence of a new collectivist mode of production which has been called the Commons-Based Peer Production (CBPP) (Benkler 2006; Rigi 2013). I argue that progressive forces can overthrow of capitalism and replace it with CBPP (Rigi 2013, 2014b). This mode of production reversing the capitalist spatial logic of space will make locality the foundation of the construction of space.

The commons-based peer production (CBPP)

CBPP is based on the production of commons by community of producers organised through voluntary cooperation. Each volunteer chooses the tasks she or he want to perform. The total communal labour distributed in this way is coordinated by a central authority which is accountable to the members of the community. Digital commons are available for free on the net. The rights to relatively scarce commons [services, forests, fisheries, etc.] are still emerging, but they might be relatively restrictive compared with rights to digital commons. The paradigmatic examples of CBPP are Linux operating system and Wikipedia. CBPP has been applied to film, design and as well to other forms of symbolic production (Benkler

2006; Soderberg 2006; Rigi 2012, 2013, 2014b. It has also been applied to the production of hardware (Rigi 2013; Kostakis *et al*. 2015).

The defining features of CBPP is its collective forms of property on the one hand and its collective mode of cooperation on the other.

The collective form of property in CBPP

Like any other form of property this collective form also consists of a bundle of juridical rights. These rights are defined by "the General Public Licence" (GPL) and its "Copyleft" clause. It was Richard Stallman who invented GPL-Copyleft in early 1980s (Stallman 2012). He did so in order to launch his Free Software (FS) Movement. Source code in software consists of human readable program. Compiler translates sources code into binary machine code. In order to modify a program one must know its source code. It is extremely difficult to detect the source code through binary machines codes (Buckman & Gay 2002:3–5). Up to 1970s when capitalists started to commercialise software engineers shared freely source code. However, in 1970s and early 1980s capitalist made source codes trade secrets and forced engineers to sign non-disclosure contracts. Copyright restrictions and non-disclosure contracts interrupted the free exchange between engineers and thereby frustrated immediate and mediated cooperation between them. This as Stallman (2002) argues, restricted engineers' freedom of expression and creativity. In order to defend and promote this freedom Stallman resigned from his post at MIT in 1984 and developed the 'Free Software' (FS). Stallman wrote an advanced operating system (GNU) and released it under the General Public License (GPL). GPL guaranteed that everyone could run the program for any purpose, study and customise it, redistribute copies of it either for free or for a reasonable price and change and improve it (Stallman 2012:20). Thus, GPL makes the knowledge to which it is attached universal commons of humanity. Copyright law gives authors the right to specify the terms of use of their work. Stallman, using this right, included the so called "Copyleft" clause in the GPL. According to this clause the whole of code/knowledge that has a component derived from a code/knowledge under GPL must also be released entirely under a GPL license (Stallman 2012:22–23). Copyleft was meant to secure the growth of commons of knowledge and prevent their privatisation. Copyleft can be understood as a dialectical negation of copyright, because, it simultaneously, preserves and abolishes copyright (Rigi 2012, 2013). The historical significance of GPL can be understood in two ways. First, it formulated for the first time in history a globally all-inclusive property right. Many people on the earth are currently unable to use this right because of a lack of access to computers and IT skills. But this is not a juridical matter, and by no means undermines the historical significance of Stallman's invention. Stallman's invention gives an important urgency to the demand that computers and IT education should be universally available for everyone. Stallman's initiative marked a major turn in the social struggle over knowledge (Stallman 2012; Soderberg 2006). It invented a property form that makes knowledge collective property of humanity. GPL-Copyleft defended "communism of science" (Merton 1979, 1996) against late-capitalist invasion. The ethos of science, Merton claimed, was communistic since science was a product of direct or indirect cooperation of scientists and was indiscriminately available to anyone who could use it. The rise of the draconian capitalist regimes of intellectual property in 1970–1980s undermined this communism by privatising commons of knowledge and science. In this context, Stallman initiative was defence of communism of science against late capitalism invasion, though Stallman did not present his project in such terms. The fruition of IT oriented scientific and technological revolutions in 1970–1980s created enormous possibilities for the expansion of the commons of science and knowledge. GPL-Copyleft provided a juridical form that could sustain the realisation of these possibilities. The neo-liberal project was launched in early 1980s. Enclosure of commons was a pillar of this project. Therefore, Stallman's project was also directly anti-neoliberal (see Rigi 2014b). This constitutes the second historical significance of Stallman's intervention.

CBPP'S mode of cooperation

GPL-Copyleft did not change the old model of scientific collective cooperation. In this model usually each individual scientist developed her or his own ideas in isolation, or a few individual members of a team together developed ideas through direct cooperation. The scientific cooperation at large scale between many scientists from different places and times was indirect and mediated through their work and not immediate communication between them (Marx 1976:199). The internet and digital technology made the large-scale cooperation between many scientific workers possible. Using this opportunity Linus Torvalds invented CBPP's mode of cooperation in 1991. Torvalds using GNU wrote the germ of an operating system and called Linux Kernel. He published it on the net under GPL and invited others to participate in developing it further. Many developers responded positively to this invitation, and after a few years, thousands of developers cooperating online created the robust Linux operating system (Raymond 2001; Weber 2004). The cooperation works as follows. There is no central division of tasks; each contributor chooses to work on problems of her own interest and solutions are published and discussed on the community's mailing list. Torvalds and his lieutenants coordinate the cooperation and have the authority to choose between alternative paths of development. However, they try to accommodate the common will of the community of participants which is shaped through democratic discussions on the mailing list. Nevertheless, if an individual or a group disagree(s) she/they can take the entire code and develop it in the direction she/they wish(es). While in Torvalds' model the coordinator is the launcher of the project in other model the coordinators are elected by the members of the cooperating community (Coleman 2013). In the most existing modes of coordination power relations in the community are asymmetrical and the coordinators may resort to authoritarian measures to impose their will. However, there is nothing that suggests that a cooperating community cannot organise itself on the basis of direct democracy, meaning that the coordinators are directly elected and disposed by the community.

The combination of digital technology, GPL-Copyleft and the digitally mediated peer to peer cooperation that produces commons constitute the new CBPP mode of production.

The application of CBPP to material production

The Linux model can be applied to all knowledge production and CBPP has been applied to the production of software, text, film, music, design, etc. (Benkler 2006). However, the fact that knowledge is the hegemonic productive force of our time combined with the digital technology will enable the generalisation CBPP to material production. Projects such as Global Village Construction Set (GVCS) for the production of ecological friendly machines (www.oepsourceecology.org.nyud.net/gvcs.php) and the RepRap project for the production of 3D printers are examples (Bowyer 2006) of the application of CBPP to the production of hardware. The rise of AI, 5G, the Internet of Things, and the expansion of the digital automation can facilitate the application of CBPP to the production of hardware. The production of hardware has two components: knowledge (R&D, design, software, etc.) and manufacturing. While the knowledge component can be produced through online peer to peer cooperation, manufacturing can be to a great extent digitally automated.

The main features of fully fledged CBPP

A fully fledged CBPP can be described as a form of advanced communism based on the following principles:

1) the action of knowledge on knowledge or on matter through the mediation of digital technology is central to production (Benkler 2006; Castells 2010; Rigi 2013).
2) the production process is collective: it is open to the participation of anyone who is able and willing to contribute and thereby develop herself or himself.
3) the contribution is voluntary and is not exchanged for money (wages). However, a contributor may feel compensated by using the common product for her own ends and enhancing her competence by learning from others. She also may receive the recognition and respect of her peers;
4) users do not pay for the products they use and are not required to contribute in order to use them. Contributors' main motivation is the materialisation of their own creativity and the creation of a commons (Himanen 2001; Weber 2004).
5) the product is universal commons, available to the member of community at large. A member is entitled to use the social wealth according to her or his needs regardless of the portion of her or his contribution to this wealth.

The spatial dialectic of CBPP

This dialectic will be decided by the particularity of the regime of production and distribution of goods and services. In a fully fledged CBPP manufacturing will be delegated to automated machines produced by automated machines minimally supervised by human beings. Thus, human being main productive activity will consist of the production of knowledge and knowledge-based services – mostly care – which cannot be automatically produced by machines or for social and ethical reasons cannot be delegated to machines. Thus, this mode of production will transcend the division between intellectual and manual labour. Surely, producer will still use their hands in some of their productive activities, but these activities will be overwhelmingly intellectual.

In this mode of production the territorial macro-social and micro-technological divisions of labour will vanish.

Currently, macro-social division of labour is an expression of the interdependence of separate branches of production that produce different types of goods. In a fully automated production the design of all goods can be stored in a computing cloud and from there sent to the producing machines. Retooled automatically the same machine can produce a range of various goods. Thus, we can assume that a limited set of automated machines can produce most material goods. The same machines retooled can produce a whole range of material goods whose designs are downloaded from the cloud. Thus, the macro-social division of labour will be replaced with the special functions of machines.

Currently, technical division of labour regulates the process of combined labour on the level of a single firm or workshop. Again this labour process will be delegated to computer-machines operating under the guidance of programs.

In short in a fully fledged and advanced CBPP the divisions between intellectual and manual labour, macro social and micro technical division of labour will all disappear. In this mode of production human beings, the nature and automated machines constitute the main forces of production. Human being will produce knowledge and knowledge-based services and machines will produce goods and services whose production are automated. Knowledge will be globally produced through peer to peer digitally enabled cooperation. The production and distribution of most services and goods will be locally organised. The multi-layered spatial scale order defined by N. Smith (1984) and D. Harvey (2003) will vanish. Instead, in this new order, we will have two main spatial scales namely the global and the local. The global scale consists of the following components: 1) the networks of collective production and transfer of knowledge; 2) networks of the production and maintenance of global infrastructure of communication and transport; 3) the networks of the production and distribution of rare strategic natural resources; and 4) the networks of global cooperation for the protection of the earth and nature.

As knowledge – a main productive force – is produced and shared globally the nature must also become global commons in two senses: first, its protection as a condition of the being of the species will become [or has already become] a global obligation of humanity as a whole; second, its strategic rare resources such as rare metals or sources of energy must be globally shared, regardless of their specific geographical locations. The global means of transport and communication must also become global commons, produced and maintained through global peer to peer cooperation. All this means that human beings will imagine themselves as the members of a unified global community of the species who appropriate and protect the nature together on the one hand and produce and share knowledge together on the other.

This global community will consist of local ones. Locality will be the building block of the constitution of the global spatial scale. I argue that locality will be the core spatial principle of a globalised CBPP based civilisation. The local is the habitat of a community, which is a sub-community of the community of the species. Thus membership of the local community is open to all living members of the community of the species. Hence, the local community is not communitarian but cosmopolitan. Likewise the local is not an isolated and insulated entity but is a nod in the global spatial networks. The local will be an active agent in the creation of the global. The global will be the space of conscious interactions, communication and exchanges between localities on the basis of equality. The global will not be a hyper-structure that dominate and alienate the local. The global scale will consist of numerous horizontally networked locals.

The spatial structure of the local

The local spaces will be urban–rural. They will be urban in the following senses: 1) they will be centres of creativity and a production of culture and art, literature, in term of creative production hey will offer more than the global city offers today; 2) they will be cosmopolitan areas where cultures and people from all corners of the planet meet and intermingle. The locals have a global perspective; and 3) as said, they will be nodes in global networks connected through communication and transport networks. They will be rural consisting of: 1) farms for the production of foods, 2) created recreational green space, and 3) wild green landscape.

While the area of these urban–rural localities may vary according to natural properties of landscape, the population must have an upper limit probably of 500,000 individuals. This limit can be kept by birth control on the one hand and the control of the inflow of people from outside to live in these localities, on the other. Inflow of people into these localities must be equal to the outflow of people from them. Any inhabitant of any locality in the planet must have the right to live in all other localities around the world. Each locality will keep a record of the excess of outflows over inflows and will admit new members to balance the difference. The admittance must be according to the principle that first in the waiting list comes first.

These urban–rural localities will be created in three ways: 1) the deconstructing of the current mega cities; 2) the expansion of some current town and villages into larger entities; and 3) the creation of new entities in the places which are favourable to the production of food and provide good ecological environment for life.

The deconstruction of the mega city and the development of villages and small towns will erase the current uneven capitalist uneven development. The deconstruction will be a very sensitive issue. It can take the following forms: 1) The inhabitants of environmentally worse neighbourhoods and the neighbourhoods of low aesthetic value will voluntary move to the expanded villages-towns or newly created urban–rural areas; 2) the nice neighbourhoods and historical sites and buildings of artistic values will be preserved; and 3) the emptied buildings whether previously living spaces, shops or offices will be levelled and the land will be transformed into green space including farmland.

This local urban–rural will consist of: a) living space; b) places of manufacturing and the production of services, and the storage for material goods; c) the centres and infrastructure of global and local communication and travel; d) farmland and pastures; e) various cultural and recreational places; and f) wild nature: forests, valleys, lakes, hills, rivers, etc.

We hypothetically suggest that two factors will dramatically reduce the places of manufacturing and the production of services, and the storage for material goods. The first is automation. First of all, the size of the future machine will be dramatically reduced compared with the size of machines today. Thus, a machine will occupy a smaller place. Second, a group of interconnected automated machines located in one place can produce most of needed material goods. Automatically, retooled the same machine can produce various goods. The space of storage of goods will also be reduced by the application of the principle of just in time production. If I need a new pair of shoes, I will check online the various designs of shoes available as global commons, choose and modify a particular design, then will send the design to automated machines and order them to produce it. I will pick them personally, or order them to be brought to my house by robots. A hospital can arrange the production of equipment it needs in the same way. Thus, space of storage will be reduced. Markets will not exist since individuals and every institution will be entitled to use the social product according to their need.

Although the storage of raw material may occupy a considerable space, the overall space devoted to the production and storage of material goods will dramatically decrease. This means that living space, farms, pastures, wild landscape, parks, theatres and other places of creative and recreational activities will dominate the urban–rural localities. Therefore, they will be green as our contemporary rural areas and simultaneously the centres of most advanced forms of innovation and creativity as the mega city is today.

Planning as self-governance and direct democracy

These urban–rural entities need governance for following purposes: 1) determining the level of material consumption; 2) the direction and scale of the physical expansion of urban–rural locality; 2) provision of necessary social services that cannot be provided safely by machines; and 3) regulating the reciprocal relations with other localities around the world including regulating inflows of new inhabitants.

Unlike the digital reproduction of knowledge, the reproduction costs of material objects, everything else being equal, is the same as the cost of their production. Therefore, their production requires the expenditure of new energy. The protection of natural environment as main priority requires a ceiling for production. This ceiling may vary depending on the level of the cleanness of the used energy. This put a limit to the magnitude of the total use-values that must be supplied in a given period. Furthermore, certain use values are naturally scarce. Thus, the distribution principle "to each according to her or his needs" must be defined and executed in a way that fits this constrain. The matching of needs, production and the protection of the environment needs planning.

The geographic and demographic expansion in term of the regulation of birth rate and the inflows of population from outside also need planning: a planning that harmonises wild nature and built space as the second nature.

There are certain services such as health care, education, hygiene, sanitation and transport which must be regularly provided. As mentioned knowledge is produced through voluntary peer-to-peer cooperation. The production of these services cannot be left to voluntary participation of the citizens, it must be planned and regulated. Certain people must be obliged to perform the tasks that are required for the production of these services. This also requires planning.

As mentioned, these urban-local entities are not isolated but must be part of global spatial networks. Knowledge is produced through global online cooperation. These localities need to regulate two physical things globally. The first is the fair distribution of the strategic natural resources, such as energy, metals, water, and wood. The second is the production and maintenance and interlinking of the equipment and infrastructure of global transport and communication. These need to be planned and coordinated by a global cooperation of urban–rural localities.

Therefore core rational for the governance will be planning. Although the planning will take place by active participant of citizens it needs to be designed and executed by experts who are supervised by the elected authorities of the urban–rural population according to the principle of direct democracy. The population can recall these authorities in any moment.

In the old socialist debates from Marx and Engels through various socialist debates in 19th and 20th centuries planning was deduced from the complexity of the technical and social divisions of labour and the need to harmonise them. Accordingly, a socialist society must calculate its needs of different goods and services for a given period, and accordingly allocate its resources – labour-force and means of production – to different branches of production and allocate the resources of each branch to individual units of production. The technical division of labour within each unit, its rhythm and productivity must be planned in concert with overall planning in such a way that the targets of the overall plan are met in pre-designated time frames.

I argue that digital automation has made the social and technical division of labour in the realm of lifeless physical production almost if not absolutely obsolete. The production of symbolic form can be organised through P2P decentred cooperation. Therefore, planning will become far less complex. It will be done through a horizontal form of governance which is based on direct democracy and by no means requires a monstrous and repressive top down bureaucracy.

Even, the major plans and their direction can be chosen by the local population as whole. The plans can be presented to public and discussed for a good while and then public can chose them by vote.

The global governance which deals with the distribution of the rare strategic raw material extracted from the earth and the regulation of the global communication and transport will emerge through democratic cooperation of localities.

Conclusions

I have presented in this chapter the emerging mode of CBPP in the realm of knowledge and speculatively indicated how it can be generalised to material production with aid of digital automation. I have also argued that the generalised CBPP can replace the capitalist mode of production. Presenting the spatial logic of this mode of production, I argue that the creation of urban–rural localities with a cosmopolitan and global orientation will constitute the core element of this logic. These localities combine the greenness of the rural with enormous cultural and artistic capability of the global city. The centrality of the locality will attend simultaneously the environmental problems and the issue of democracy. Planning and direct democracy are the means of the organisation of economic, ecological, social and cultural life within these localities.

The question arises: to what extent is all this viable? Haven't I concocted an idle fantasy? The question is pressing in the light of fact that the current actual spatial development is oriented toward of the creation of the capitalist mega city. My project is contrafactual and counter intuitive. A closer inspection, however, shows that my project is potentially embedded in the science and technology on the one hand and the desires of progressive social movements on the other.

I acknowledge that what I have presented is an abstraction derived from a certain scientific and technological developments that have spurred the emergence of an embryonic mode of production, namely CBPP. As an abstraction it is a simplification and suffers from all defections of abstraction. However, it is not meant to offer an actual design for the production of future urban–rural localities and their global connections in the future. It is neither a claim that such localities will ever emerge. It is not concerned with space as a thing but tries to identify a new principle of the process of the production of space – for the distinction between process and thing in relation to space production, see David Harvey (1997). The form of the actual processes and the final products of these processes namely the actual urban–rural localities of the future, if they will come about at all, will be far more complex and I would not venture to speculate about them.

With these caveats in mind, I argue that the digital technology, automation and CBPP make the translation of the principle described here into processes of production of space viable. The hegemony of tendency towards locality and direct democracy in contemporary social movements, especially the environmental movements (Castells 2001) may strengthen the viability of this principle.

Another fact that apparently speaks against the viability of this model-principle is that the strategic means of production including the infrastructures of communication and transport on the one hand and land on the other are privately owned. In the model these means of productions are commons. I acknowledge that the viability of this principle is conditioned upon the abolishing of private property in strategic means of production, especially land which requires a series of social revolutions. Such social revolutions can be conducted by social movements. I by no means suggest that such revolutions, their success and the coming of Board Certified Psychiatric Pharmacist (BCPP) are inevitable evolutionary processes. All depends on social movements, their choices and strengths and capability for mobilising the majority of people for materialising their goals. As ever, making of history is contingent upon political and social struggle.

References

Aristotle, 1885, *Politics*, trans. B. Jowett, Clarendon Press, Oxford.

Aristotle, 1996, *The Nicomachean ethics*, trans. H. Rackham, Wordsworth, Hertfordshire.

Benkler, Y., 2006, *The wealth of networks*, Yale University Press, New Haven.

Bowyer, A., 2006, *The self-replicating rapid prototype: Manufacturing for the masses*, http://reprap.org/wiki/PhilosophyPage.

Buckman, R.E. & Guy, J., 2002, 'A note on software', in S., Richard (ed.), *Free software, free society*, pp. 3–5, GNU Press, Boston, MA.

Castells, M., 2001, *The power of identity*, Blackwell, Oxford.

Castells, M., 2010, *The rise of network society*, Wiley-Blackwell, Oxford.

Coleman, G., 2013, *Coding freedom: ethics and aesthetics of hacking*, Princeton University Press, Princeton.

Engels, F., 1968, 'The origin of the family, private property and state', in *Marx and Engels selected works*, pp. 468–593, Progress Publishers, Moscow.

Engels, F., 2000, *The origin of the family, private property and the state*, Marxist.org.

Harvey, D., 1982, *Limits to capital*, Blackwell, Oxford.

Harvey, D., 1997, *Justice, nature and the geography of difference*, Blackwell, Oxford.

Harvey, D., 2003, *New imperialism*, Oxford University Press, Oxford.

Harvey, D., 2012, *Rebel cities: from the right to the city to urban revolution*, Verso, New York.

Himanen, P., 2001, *The hacker Ethic: a radical approach to the philosophy of business*, Random House, New York.

Kostakis, V. *et al.*, 2015, 'Design global, manufacture local: Exploring the contours of an emerging productive model', *Futures* 73, 126–135.

Lefebvre, H., 2003, *The urban revolution*, University of Minnesota Press, Minneapolis.

Merton, R., 1979, *The sociology of science: Theoretical and empirical investigations*, Chicago University Press, Chicago.

Merton, R., 1996, *On social structure and science*, Chicago University Press, Chicago.

Mandel, E., 1975, *Late capitalism*. Verso, London.

Marx, K., 1976, *Capital*, vol, 1, Penguin Books, London.

Marx, K. 1981, *Capital*, vol. 3, Penguin Books, London.

O'Neil, M, 2009, *Cyber Chiefs: Autonomy and Power in Online Tribes*, Pluto Press, London.

Raymond, E., 2001, *The cathedral and bazar*, O'Reilly, Sebastopol.

Rigi, J., 2012, 'Peer production as an alternative to capitalism: A new communist horizon', *Journal of Peer Production* 1(1) http://www.indymedia.org.uk/en/2012/08/498544.html.

Rigi, J., 2013, 'Peer production and Marxian communism: Contours of new emerging mode of production', *Capital & Class* 37(3), 397–416.

Rigi, J., 2014a, 'Foundations of a Marxist theory of the political economy of information: Trade secrets and intellectual property, and the production of surplus-value and extraction of rent', *Triple C* 12(2), 909–936.

Rigi, J., 2014b, 'The coming revolution of peer production and revolutionary cooperative. A response to Michel Bauwens, Vasilis Kostakis and Stefan Meretz', *Triple C* 12(1), 390–404.

Rigi, J., & Prey, R., 2015, 'Value, rent, and the political economy of social media', *The Information Society* 31(5), 392–406.

Smith, N., 1984, *Uneven development: Nature, capital and the production of space*, Blackwell, Oxford.

Sassen, S., 2011, *Cities in a world economy*. Pine Forge Press, Thousand Oaks, CA.

Soderberg, J., 2006, *Hacker capitalism*. Routledge, London.

Stallman, R., 2002, *Free software, free society*, GNU Press, Boston, MA.

Turnbull, C., 1961, *Forest people*. Simon & Schuster, New York.

Weber, S., 2004, *The success of open source*, Harvard University Press, Cambridge, MA.

Wood, E.M., 2001, *Empire of capital*, Verso, London.

Woolf, E., 2010, *Europe and people without history*, University of California Press, Berkeley.

Index

Note: figures and tables are denoted with *italicized* and **bold** page numbers, respectively.

Printed in the United States
by Baker & Taylor Publisher Services

Printed in the United States
by Baker & Taylor Publisher Services